Studies in Computational Intelligence 429

For further volumes:
http://www.springer.com/series/7092

Roger Lee (Ed.)

Computer and Information Science 2012

Editor
Roger Lee
Software Engineering & Information Technology Institute
Computer Science Department
Central Michigan University
Mt. Pleasant Michigan
USA

ISSN 1860-949X e-ISSN 1860-9503
ISBN 978-3-642-42934-7 ISBN 978-3-642-30454-5 (eBook)
DOI 10.1007/978-3-642-30454-5
Springer Heidelberg New York Dordrecht London

Softcover reprint of the hardcover 1st edition 2012

Printed on acid-free paper

Springer is part of Springer Science+Business Media (www.springer.com)

Preface

The purpose of the 11th International Conference on Computer and Information Science(ICIS 2012) held on May 30 – June 1, 2012 in Shanghai, China was to bring together researchers and scientists, businessmen and entrepreneurs, teachers and students to discuss the numerous fields of computer science, and to share ideas and information in a meaningful way. Our conference officers selected the best 15 papers from those papers accepted for presentation at the conference in order to publish them in this volume. The papers were chosen based on review scores submitted by members of the program committee, and underwent further rounds of rigorous review.

In Chapter 1, In this paper, we address a challenging task of automat generation of UML class models. In conventional CASE tools, the export facility does not export the graphical information that explains the way UML class elements (such as classes, associations, etc) are represented and laid out in diagrams. We address them problem by presenting a novel approach for automatic generation of UML class diagrams using the Binary Space Partitioning (BSP) tree data structure. A BSP tree captures the spatial layout and spatial relations in objects in a UML class model drawn on a 2-D plane. Once the information of a UML model is captured in a BSP tree, the same diagram can be re-generated by efficient partitioning of space (i.e. regions) without any collision. After drawing UML classes, the associations, aggregations and generalisations are also drawn between the classes. The presented approach is also implemented in VB.NET as a proof of concept. The contribution does not only assist in diagram interchange but also improved software modeling.

In Chapter 2, The Semantic Web is recognized as the next generation web which aims at the automation, integration and reuse of data across different internet applications. To better understand and utilize the Semantic Web, the W3C adopted standards and tools such as the Resource Description Framework (RDF) and Web Ontology Language (OWL). Management of large amounts of semantic data, stored in semantic models, are required within almost every semantic web application environment, thus motivating the design of specific repositories in order to store and operate semantic models. However, most semantic storage systems based on a relational database support both a monolithic schema with a single table that stores all statements. Since the Semantic Web systems use OWL ontologies as set of RDF triples (which are not composed of hierarchical knowledge between class and property) rather than complex OWL data models, they are not compliant with the more advanced systems. Further, an expected reduced system performance can be observed due to the large amount of semantic data managed in a single storage model. We propose improving the storage model of OWL by changing the management of OWL data and designing an efficient new relational database layout to store an OWL knowledge base, focused on the OWL 2 DL language relational. Upon evaluation, our storage system shows an improvement in the performance and management.

In Chapter 3, This paper presents the DPF Workbench, a diagrammatic tool for domain specific modelling. The tool is an implementation of the basic ideas from the Diagram Predicate Framework (DPF), which provides a graph based formalisation of (meta)modelling and model transformations. The DPFWorkbench consists of a specification editor and a signature editor and offers fully diagrammatic specification of domain-specific modelling languages. The specification editor supports development of metamodelling hierarchies with an arbitrary number of metalevels; that is, each model can be used as a metamodel for the level below. The workbench also facilitates the automatic generation of domain-specific specification editors out of these metamodels. Furthermore, the conformance relations between adjacent metalevels are dynamically checked by the use of typing morphisms and constraint validators. The signature editor is a new component that extends the DPF Workbench with functionality for dynamic definition of predicates. The syntax of the predicates are defined by a shape graph and a graphical icon, and their semantics are defined by validators. Those predicates are used to add constrains on the underlying graph. The features of the DPF Workbench are illustrated by a running example presenting a metamodelling hierarchy for workflow modelling in the health care domain.

In Chapter 4, With the increasing number of web services deployed to the world wide web these days, discovering, recommending, and invoking web services to fulfil the specific functional and preferential requirements of a service user has become a very complex and time consuming activity. Accordingly, there is a pressing need to develop intelligent web service discovery and recommendation mechanisms to improve the efficiency and effectiveness of service-oriented systems. The growing interests in semantic web services has highlighted the advantages of applying formal knowledge representation and reasoning models to raise the level of autonomy and intelligence in human-to-machine and machine-to-machine interactions. Although classical logics such as description logic underpinning the development of OWL has been explored for services discovery, services choreography, services enactment, and services contracting, the non-monotonicity in web service discovery and recommendation is rarely examined. The main contribution of this paper is the development of a belief revision logic based service recommendation agent to address the non-monotonicity issue of service recommendation. Our initial experiment based on real-world web service recommendation scenarios reveals that the proposed logical model for service recommendation agent is effective. To the best of our knowledge, the research presented in this paper represents the first successful attempt of applying belief revision logic to build adaptive service recommendation agents.

In Chapter 5, Currently, signature-based network intrusion detection systems (NIDSs) have been widely deployed in various organizations such as universities and companies aiming to identify and detect all kinds of network attacks. However, the big suffering problem is that signature matching in these detection systems is too expensive to their performance in which the cost is at least linear to the size of an input string and the CPU occupancy rate can reach more than 80 percent in the worst case. This problem is a key limiting factor to encumber higher performance of a signature-based NIDS under a large-scale network. In this paper, we developed an exclusive signature matching scheme based on single character frequency to improve the efficiency of traditional signature matching. In particular, our scheme calculates the single character frequency from both stored and matched NIDS signatures. In terms of a decision algorithm, our scheme can adaptively choose the most appropriate character for conducting the exclusive signature matching in distinct network contexts. In the experiment, we implemented our scheme in a constructed network environment and the experimental results show that our scheme offers overall improvements in signature matching.

In Chapter 6, Data-intensive scientific workflow based on Hadoop needs huge data transfer and storage. Aiming at this problem, on the environment of an executing computer cluster which has limited computing resources, this paper adopts the way of data prefetching to hide the overhead caused by data search and transfer and reduce the delays of data access. Prefetching algorithm for data-intensive scientific workflow based on the consideration of available computing resources is proposed. Experimental results indicate that the algorithm consumes less response time and raises the efficiency.

In Chapter 7, Fingerprint orientation field estimation is an important processing step in a fingerprint identification system. Orientation field shows a fingerprint's whole pattern and globally depicts the basic shape, structure and direction. Therefore, how to exactly estimate the orientation is important. Generally, the orientation images are computed by gradient-based approach, and then smoothed by other algorithms. In this paper we propose a new method, which is based on nonnegative matrix factorization (NMF) algorithm, to initialize the fingerprint orientation field instead of the gradient-based approach. Experiments on small blocks of fingerprints prove that the proposed algorithm is feasible. Experiments on fingerprint database show that the algorithm has a better performance than gradient-based approach does.

In Chapter 8, We present LCT_{SG}, an LSC (Live Sequence Chart) consistency testing system, which takes LSCs and symbolic grammars as inputs and performs an automated LSC simulation for consistency testing. A symbolic context-free grammar is used to systematically enumerate continuous inputs for LSCs, where symbolic terminals and domains are introduced to hide the complexity of different inputs which have common syntactic structures as well as similar expected system behaviors. Our symbolic grammars allow a symbolic terminal to be passed as a parameter of a production rule, thus extending context-free grammars with context-sensitivity on symbolic terminals. Constraints on symbolic terminals may be collected and processed dynamically along the simulation to properly decompose their symbolic domains for branched testing. The LCT_{SG} system further provides either a state transition graph or a failure trace to justify the consistency testing results. The justification result may be used to evolve the symbolic grammar for refined test generation.

In Chapter 9, Mining evolving behavior over multi-dimensional structures is increasingly critical for planning tasks. On one hand, well-studied techniques to mine temporal structures are hardly applicable to multidimensional data. This is a result of the arbitrary-high temporal sparsity of these structures and of their attribute-multiplicity. On the other hand, multi-label classifications over denormalized data do not consider temporal dependencies among attributes. This work reviews the problem of long-term classification over multidimensional structures to solve planning tasks. For this purpose, firstly, it presents an essential formalization and evaluation method for this novel problem. Finally, it extensively overviews potential relevant contributions from different research streams.

In Chapter 10, A growing challenge in data mining is the ability to deal with complex, voluminous and dynamic data. In many real world applications, complex data is not only organized in multiple database tables, but it is also continuously and endlessly arriving in the form of streams. Although there are some algorithms for mining multiple relations, as well as a lot more algorithms to mine data streams, very few combine the multi-relational case with the data streams case. In this paper we describe a new algorithm, Star FP-Stream, for finding frequent patterns in multi-relational data streams following a star schema. Experiments in the emphAdventureWorks data warehouse show that Star FP-Stream is accurate and performs better than the equivalent algorithm, FP-Streaming, for mining patterns in a single data stream.

In Chapter 11, Recently fast innovation of Internet technology causes lot of application to change into mobile application and the technology trends of communication equipment are changed from mono-function to multi-functioned system. These trends are part of changes which is caused by ubiquitous world and it is just beginning of huge waves which is required to fit and change under the ubiquitous environments. In this paper, we focused on the Design and Implementation of Component Objects that can be communicated effectively among various types of clients under the Heterogeneous Client Server Environments and Material Management System was chosen as the target of application. The key point to do that kind of affair is using component objects for the enforcement of reusability and inter-operability among and using XML mobile services that can communicate thru systems. Thus the Components proposed in this paper could be reused effectively in case of developing similar applications.

In Chapter 12, In recent years, the data-driven peer-to-peer streaming systems have been extensive deployed in Internet. In these systems, the transfer of media blocks among nodes involves two scheduling issues: each node should request the streaming blocks of interest from its peers (i.e. block request scheduling), on the other hand, it also should decide how to satisfy the requests received from its peers (i.e. block delivery scheduling) given its bandwidth limitations. Intuitively, these two scheduling issues are critical to the performance of data-driven streaming systems. However, most of the work in the literature focused on the block request scheduling issue, and very few concentrated on the latter. Consequently, the performance of the system may be affected seriously due to an unsophisticated scheduling strategy. In this paper, we analytically study the block delivery scheduling problem and model it as an optimization problem based on the satisfaction degrees of nodes and the playback performance of the system. We then propose a scheduling strategy and prove the optimality of the strategy to the optimization problem. Lastly, we illustrate the effectiveness of the proposed strategy by extensive simulation.

In Chapter 13, One common approach or framework of self-adaptive software is to in-corporate a control loop that monitoring, analyzing, deciding and executing over a target system using predefined rules and policies. Unfortunately, policies or utilities in such approaches and frameworks are statically and manually defined. The empirical adaptation policies and utility profiles cannot change with environment thus cannot make robust and assurance decisions. Various efficiency improvements have been introduced to online evolution of self-adaptive software itselfhowever, there is no framework with policy evolution in policy-based self-adaptive software such as Rainbow. Our approach, embodied in a system called IDES(Intelligent Decision System) uses reinforcement learning to provides an architecture based self-adaptive framework. We associate each policy with a preference value.During the running time the system automatically assesses system utilities and use reinforcement learning to update policy preference. We evaluate our approach and framework by an example system for bank dispatching. The experiment results reveal the intelligence and reactiveness of our approach and framework.

In Chapter 14, This study focuses on the visual representation of mathematical proofs for facilitating learners' understanding. Proofs are represented by a system of sequent calculus. In this paper, the authors discuss SequentML, an originally designed XML (Extensible Markup Language) vocabulary for the description of sequent calculus, and the visualization of mathematical proofs by using this vocabulary.

In Chapter 15, Incremental construction of fuzzy rule-based classifiers is studied in this paper. It is assumed that not all training patterns are given a priori for training classifiers, but are gradually made available over time. It is also assumed the previously available training patterns cannot be used in the following time steps. Thus fuzzy rule-based classifiers should be constructed by updating already constructed classifiers using the available training patterns at each time step. Incremental methods are proposed for this type of pattern classification problems. A series of computational experiments are conducted in order to examine the performance of the proposed incremental construction methods of fuzzy rule-based classifiers using a simple artificial pattern classification problem.

It is our sincere hope that this volume provides stimulation and inspiration, and that it will be used as a foundation for works yet to come.

May 2012 Roger Lee

Contents

List of Contributors

Abdullah Alamri
RMIT University, Australia
E-mail: abdullah.alamri@rmit.edu.au

Cláudia Antunes
IST-UTL, Portugal
E-mail: claudia.antunes@ist.utl.pt

Imran Sarwar Bajwa
University of Birmingham, U.K.
E-mail: i.s.bajwa@cs.bham.ac.uk

Andrzej Bargiela
Osaka Prefecture University, Japan
E-mail:
andrzej.bargiela@nottingham.ac.uk

Peter Bertok
RMIT University, Australia
E-mail: peter.bertok@rmit.edu.au

Gaozhao Chen
Shanghai University, China
E-mail: chengaozhao2008@163.com

Zhi Chen
Shenzhen Institute of Information Technology, China
E-mail: mufchen@gmail.com

Yunwen Ge
Shanghai University, China
E-mail: geyunwen@shu.edu.cn

Rongrong Gu
Shanghai University, China
E-mail: gurong_rong@126.com

Xiaodong Gu
Nanjing University, China
E-mail: guxiaodong1987@126.com

Hai-Feng Guo
University of Nebraska at Omaha, USA
E-mail: haifengguo@unomaha.edu

Tiande Guo
Graduate University of Chinese Academy of Sciences Beijing, China
E-mail: tdguo@gucas.ac.cn

Kashif Hameed
The Islamia University of Bahawalpur, Pakistan
E-mail: gentle_kashif@yahoo.com

Congying Han
Graduate University of Chinese Academy of Sciences Beijing, China
E-mail: hancy@gucas.ac.cn

Yang Hao
Graduate University of Chinese Academy of Sciences Beijing, China
E-mail: haoyang10@mails.gucas.ac.cn

Rui Henriques
IST-UTL, Portugal
E-mail: rmch@ist.utl.pt

Guowei Huang
Shenzhen Institute of Information Technology, China
E-mail: huanggw@sziit.edu.cn

Haeng-Kon Kim
Catholic University of Daegu, Korea
E-mail: hangkon@cu.ac.kr

Lam-for Kwok
City University of Hong Kong,
Hong Kong SAR
E-mail: cslfkwok@cityu.edu.hk

Yngve Lamo
Bergen University College
E-mail: yla@hib.no

Raymond Y.K. Lau
City University of Hong Kong,
Hong Kong SAR
E-mail: raylau@cityu.edu.hk

Roger Y. Lee
Central Michigan University, USA
E-mail: lee1ry@cmich.edu

Wenjuan Li
City University of Hong Kong,
Hong Kong SAR
E-mail: wenjuan.anastatia@gmail.com

Wendy MacCaull
St. Francis Xavier University
E-mail: wmaccaul@stfx.ca

Florian Mantz
Bergen University College
E-mail: fma@hib.no

Yuxin Meng
City University of Hong Kong,
Hong Kong SAR
E-mail: ymeng8@student.cityu.edu.hk

Yoshinori Miyazaki
Shizuoka University, Japan
E-mail: yoshi@inf.shizuoka.ac.jp

Tomoharu Nakashima
Osaka Prefecture University, Japan
E-mail: nakashi@cs.osakafu-u.ac.jp

Adrian Rutle
St. Francis Xavier University
E-mail: arutle@stfx.ca

Guangqi Shao
Graduate University of Chinese
Academy of Sciences Beijing, China
E-mail: guangqi-002@163.com

Andreia Silva
IST-UTL, Portugal
E-mail: andreia.silva@ist.utl.pt

Cuicui Song
Shanghai University, China
E-mail: scui1126@163.com

Long Song
City University of Hong Kong,
Hong Kong SAR
E-mail: longsong@student.cityu.edu.hk

Mahadevan Subramaniam
University of Nebraska at Omaha,
USA
E-mail: msubramaniam@unomaha.edu

Takeshi Sumitani
Osaka Prefecture University, Japan
E-mail:
takeshi.sumitani@ci.cs.osakafu-u.ac.jp

Xiaoliang Wang
Bergen University College
E-mail: xwa@hib.no

Takayuki Watabe
Shizuoka University, Japan
E-mail: gs11055@s.inf.shizuoka.ac.jp

Shaochun Wu
Shanghai University, China
E-mail: scwu@shu.edu.cn

Lingyu Xu
Shanghai University, China
E-mail: xly@shu.edu.cn

Yongquan Xu
Shanghai University, China
E-mail: xuyongquan54321@126.com

Generating Class Models Using Binary Space Partition Algorithm

Kashif Hameed and Imran Sarwar Bajwa

Abstract. In this paper, we address a challenging task of automat generation of UML class models. In conventional CASE tools, the export facility does not export the graphical information that explains the way UML class elements (such as classes, associations, etc) are represented and laid out in diagrams. We address them problem by presenting a novel approach for automatic generation of UML class diagrams using the Binary Space Partitioning (BSP) tree data structure. A BSP tree captures the spatial layout and spatial relations in objects in a UML class model drawn on a 2-D plane. Once the information of a UML model is captured in a BSP tree, the same diagram can be re-generated by efficient partitioning of space (i.e. regions) without any collision. After drawing UML classes, the associations, aggregations and generalisations are also drawn between the classes. The presented approach is also implemented in VB.NET as a proof of concept. The contribution does not only assist in diagram interchange but also improved software modeling.

Keywords: UML Class Models, Binary Space Partition Tree, XMI, XML.

1 Introduction

Since the emergence of the Object Oriented Modeling (OOM), the software design patterns have been improved to assist the programmers to address the complexity

Kashif Hameed
Department of Computer Science, The Islamia University of Bahawalpur,
63100, Bahawalpur, Pakistan
e-mail: gentle_kashif@yahoo.com

Imran Sarwar Bajwa
School of Computer Science, University of Birmingham, B15 2TT, Birmingham, UK
e-mail: i.s.bajwa@cs.bham.ac.uk

R. Lee (Ed.): Computer and Information Science 2012, SCI 429, pp. 1–13.
springerlink.com

of a problem domain in a better way. The OOM suggests handling a problem as a set of related and interacting Objects instead of considering as a set of functions that can be performed. A key feature of OOM is re-usability of same piece of information. In OOM, Unified Modeling Language (UML) based graphical notation is used to represent a model or a schema. There are various CASE tools such as Rational Rose, USE, Enterprise Architect, ArgoUML, Altova, Smart Draw, MS Visio, etc. All these tools provide a facility to export metadata of a UML class model using XML Metadata Interchange (XMI) [1] representation. XMI provides a standardized representation of the metadata information in a UML model so that it may be exchanged across many industries and operating environments, different UML tools and other tools that are capable of using XMI. XMI uses XML (Extensible Markup Language) representation to transport information that is highly internally referential.

Considering the undisputable significance of XMI, most of the CASE tools facilitate a user to import and export UML class models in XMI format. However, there is a limitation in current implementation of import/export facility provided by all the CASE tools that only metadata (names of classes, attributes, methods, associations, etc) of a UML class model can be exported or imported. Due to this limitation, a complete UML class diagram can not be interchanged among various CASE tools and it means that UML class diagram dawn in a CASE tool cannot be reused in other CASE tools. This limitation nullifies the basic principal of OOM that is reusability of information. A principal reason of this limitation is that the graphical visualization of a model from XMI is not possible since XMI has no graphic information [19] such as where to draw a model element and how to avoid overlapping of the model elements. In absence of this facility, the real power of XMI remains un-explored.

To address this challenging task, we present an approach for automatic generation of UML class diagrams from XMI using the Binary Space Partitioning (BSP) tree [20]. The BSP trees are typically used in the field of computer graphics to capture graphical information of a 2-D diagram. A BSP tree captures the spatial layout and spatial relations in objects in a UML class model drawn on a 2-D plane. Once the information of a UML model is captured in a BSP tree, the same diagram can be re-generated by efficient partitioning of space (i.e. regions) without any collision. After drawing UML classes, the associations, aggregations and generalisations are also drawn between the classes. We have also implemented the presented approach in VB.NET as a proof of concept and we have also solved a case study to validate the performance of our approach.

Rest of the paper is ordered into various sections: Section 2 highlights the work related to the presented research. Section 3 explains algorithm used for space portioning and its implementation is explained in Section 3. Section 4 presents a case study that manifests the potential of the presented approach in real time software engineering. The paper is concluded with the possible future enhancements.

2 Related Work

Due to limitation of XMI, diagram interchange is not possible in real meanings. Not very much work has been presented to address the limitation of XMI. To address the same issue, Marko [19] proposed some major changes in XMI metamodel to store graphical information of a UML model. However, we think that this is not a good solution due to the reason that the XMI metamodel will become far complex with addition of graphical information and the handling of XMI will become more difficult. The gap in presented work to-date was the real motivation for the presented work.

Similarly, to generate UML diagrams from XML data, Mikael, et al [9] presented an algorithm in 2001. The presented work was primarily focusing on the use of the generated diagrams for the conceptual design of (virtual) a data warehouses with respect to the web data. Another major intention of the research was to support the On-Line Analytical Processing (OLAP) based on web data. However, according to best of our knowledge the generation of UML class diagrams from XMI representation is a novel idea and no approach/tool has been presented to date for this transformation.

Some work has also been presented in automated UML class diagram generation from a natural language (NL) available in [10], [13-18], [26-28]. These approaches extract UML class elements from NL specification of software requirements and generate UML class diagrams. Such work was also the motivation of the presented approach.

To our best of knowledge, no work has been presented yet for generation of UML graphical models from the XMI presentation. Missing support for XMI to UML models transformation breaches a gap in XMI and UML. To fill this gap, there is need of an approach for automated transformation of XMI to UML to make XMI more effective and powerful.

3 Generating UML Class Diagrams from BSP Tree

The presented approach works in two steps. In first step, a BSP tree is generated from XMI representation of a UML class model. In second step, BSP tree is traversed and the UML diagram is generated. The presented approach can process XMI 2.1 format. We have chosen XMI 2.1 format as all major tools such as ArgoUML, Enterprise Architect, USE, Altova UModel support XMI 2.1 format. User provides the input in the form of XMI (.xml) file using the interface provided in the typical software system. Fig. 1 represents the overview of all steps involved in XMI to UML class diagram transformation. Following are details of the involved steps in generation of UML class diagrams from XMI 2.1 using BSP tree data structure.

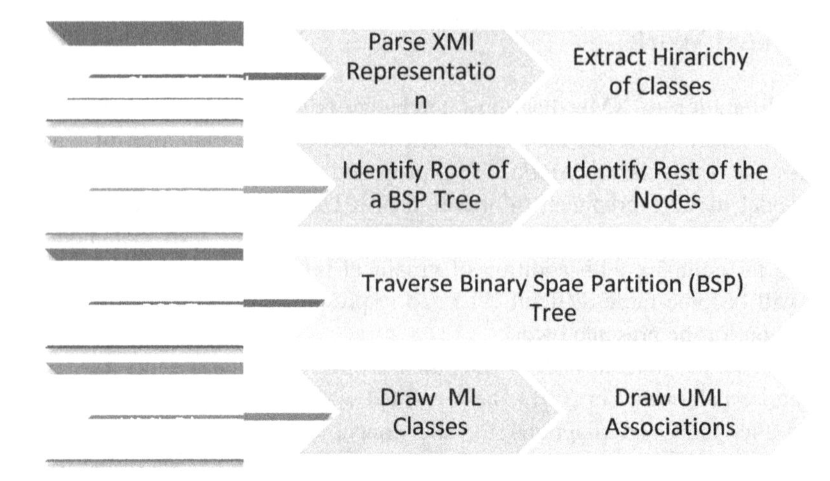

Fig. 1 Framework for Generation of UML class diagram from XMI

3.1 Parse XMI Representation

This module parses the given input (XMI representation of UML class model) using our XMI parser developed in VB.NET.

The XMI parser parses an XMI file with respect to target UML model by storing the data it in memory. For parsing XMI with VB.Net we used the `System.XML` namespace. The XML classes in the `System.Xml` namespace provide a comprehensive and integrated set of classes, allows working with XMI representation. These XML classes support parsing and writing XML, editing XML data in memory, and data validation. There are different classes available to read and write XML document in .Net Framework such as `XmlReader`, `XmlNodeReader`, `XmlTextReader`, and `XmlValidatingReader`.

A XMI parser is developed and used to parse XMI file to identify and extract class names, class attributes, class functions, association names, generalizations names, etc. While parsing an XMI presentation, XMI parser checks the syntax for well-formedness, and reports any violations (reportable errors). It helps to extract classes, objects and related tags.

The output of this phase is set of classes, their attributes, and methods, and all type of relationships such as associations, aggregations and generalizations.

3.2 Identifying Hierarchy of Classes

Once the information is extracted from XMI representation, next step is to identify relationships among multiple (two or more than two) classes and maps the associations and generalizations in classes and objects. Primary objective of this step is to extract possible hierarchy among all the classes. Such type of information helps in generating a BSP tree. Following rules are used to identify hierarchy among classes:

i. If a class *A* is associated to class *B*, class *B* will become child of class *A*.
ii. If a class *A* is generalization of another class *B*, class *B* will become child of *A* as *B* will be inheriting all features of *B*.
iii. If a class *A* aggregates another class *B*, class *B* will become child of *A*.
iv. If there is a class that has no relationship to other classes or there is numeration, which will be considered as leaves of a tree.

3.3 Generating a BSP Tree

We need to generate a BSP tree that can spatially distribute all classes into a diagram. By using the information (such as classes, associations, etc) extracted in section 3.1 and the hierarchal information identified in section 3.2, a BSP tree is constructed in this step. Then each class becomes a node of the root on the basis of the identified hierarchy. This process is recursively repeated in each half-space until every class has been incorporated into the tree. We have used the following algorithm to generate a BSP tree:

Step 1: Get a list of all the classes.

Step 2: Select a middle class *m*. Middle class *m* is selected as if total number of classes *n* is odd then $m = {}^{n}/_{2}$ else $m = {}^{n}/_{2} + 1$

Step 3: Put all of the classes before the class m, in XMI file, into a list *l* for left side of the BSP tree.

Step 4: Put all of the classes after the class m, in XMI file, into a list *r* for left side of the BSP tree.

Step 5: The class *m* simply becomes the root of the BSP tree. Thereafter, the classes in list *l* are placed in the BSP tree to the left of its parent class while the classes in list *r* are placed in the BSP tree to the right of its parent class.

Step 6: If list *l* and *r* has more elements, go to step2. Recursively repeat the process for both lists *l* and *r* until the last item in each list.

Once a BSP tree is generated, it is ready to be traversed and generate a UML class diagram. The traversal details of a BSP tree are given in next section.

3.4 Traversing the BSP Tree

To generate the UML class diagrams, we need to traverse the BSP tree, first. The tree is typically traversed in linear time from an arbitrary viewpoint. However, we are not drawing the UML model in a particular perspective of user's view. Hence, we do not consider the view point parameter here. Following algorithm was used for BSP tree traversal:

```
Function traverse(BSP_tree){
      if (BSP_tree != null){
         if (positive_side_of(root(BSP_tree))
            traverse(positive_branch(BSP_tree));
            display_triangle(root(BSP_tree));
            traverse(negative_branch(BSP_tree));
         else
            traverse(negative_branch(BSP_tree));
            display_triangle(root(BSP_tree));
            traverse(positive_branch(BSP_tree);
```

Fig. 2 Algorithm used for traversal of a BSP tree

In computer graphics, a BSP tree is in-order traversed. The process of in-order traversal recursively continues until the last node of the tree is traversed. In the case of a 2-*d* space tree, a modified in-order traversal (see Fig. 2) yields in a depth-sorted ordering of all rectangles (classes) in the space. Using the in-order traversal, either a back-to-front or front-to-back ordering of the triangles (classes) can be drawn. The back-to-front or front-to-back ordering is a matter of concern in the scenes where there are overlapping objects. However, in case of a UML class model, all objects (classes) in a scene are non-overlapping; the ordering of drawing does not matter.

3.5 Drawing UML Class Model

In this phase, first the extracted information from the previous module is used to draw the UML class diagrams. First of all the space is vertically divided from center. The class at root can be drawn at any side of the vertical division. Then each side (left & right) are horizontally and vertically drawn for classes represented by the child nodes. This process continues recursively until the last class is drawn.

Two physically draw class diagrams, a diagram generator was written in VB.NET using 2D Graphics library. Various graphics methods in GDI+ such as the DrawLine method draws a line between two points specified by a pair coordinates. DrawLines draws a series of lines using an array of points. We have also used GDI+ Graphics paths to redraw certain graphics items. The Graphics paths are used to handle multiple graphics items, including lines, rectangles, images, and text associated with a surface but we need to redraw only the rectangles. We can create a graphics path with all class diagrams (three rectangles) and just redraw that path, instead of the entire surface.

Once the rectangles for each class are drawn, each rectangle is filled with their respective attributes and methods. Then, the associations, aggregations and generalizations are drawn among the respective rectangles representing particular classes. The last step is to label the generated class diagrams with additional supportive information to make the class model more expressive.

4 Tool Support

The approach presented in section 4.2 was implemented in VB.NET as a proof of concept. A prototype tool was generated that can not only read XMI representation but also maps XMI information to a UML diagram. In XMI to UML class model mapping, a challenge was the implementation of BSP algorithm from generating UML classes as this has not been done before. The implementation of XMI2UML prototype tool consists of three key modules as below:

1. XMI Parser
2. BSP Tree Handler
3. Diagram Generator

Following is an overview of all these three modules those are counterpart of the XMI2UML prototype system. A screenshot of the implementation of XMI2UML tool in VB.NET is shown in Figure 3:

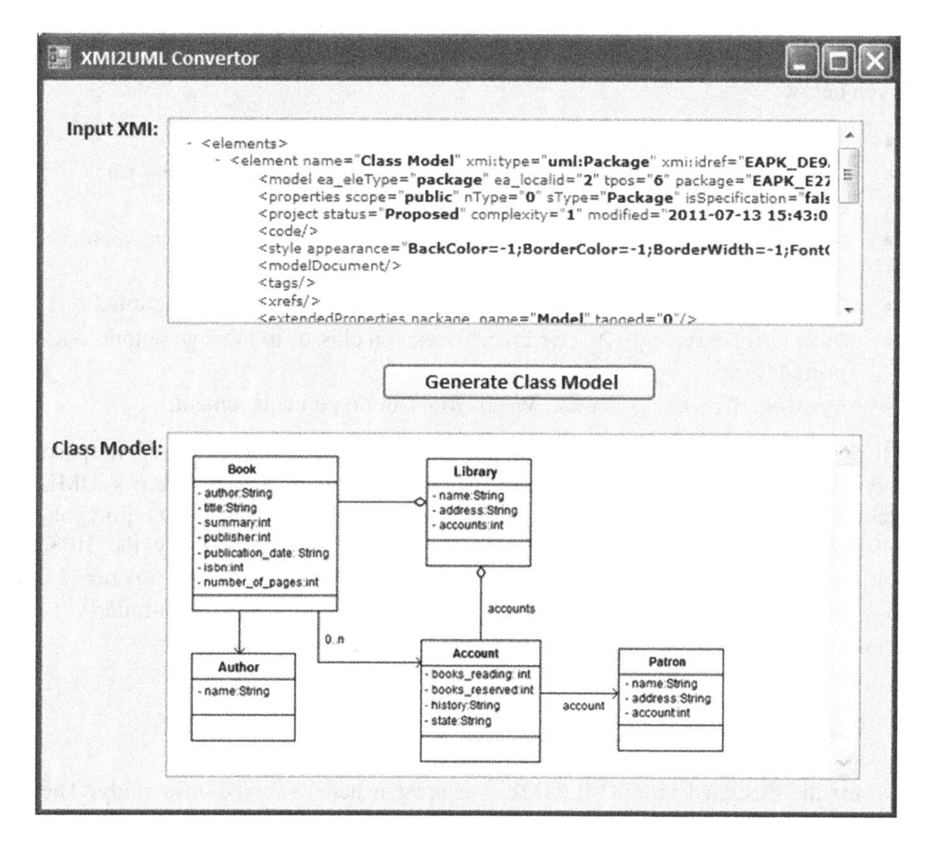

Fig. 3 Screenshot of XMI2UML prototype tool

4.1 XMI Parser

This module receives an XMI file and extracts all the UML class related information such as classes, attributes, methods, associations, etc. To read XMI representation arious classes are available in .Net Framework such as `XmlReader`, `XmlNodeReader`, `XmlTextReader`, and `XmlValidatingReader`. By the help of these classes, our XMI parser parses XMI file and extracts the required information.

4.2 BSP Tree Handler

The second module, BSP tree handler, receives output of the XMI parser and generates a BSP tree by using the algorithm discussed in section 4.2.3. In implementation of BSP tree generation algorithm, we used VB.NET ArrayList data structure.

4.3 Diagram Generator

The third and last module is diagram generator. To draw various shapes, we have used the .NET Framework Class Library that involves a whole host of classes given below:

- `System.Drawing`: Provides basic graphics functionality.
- `System.Drawing.Design`: Focuses on providing functionality for extending the design time environment.
- `System.Drawing.Drawing2D`: Provides two-dimensional and vector graphics classes and methods.
- `System.Drawing.Imaging`: Exposes advanced imaging functionality.
- `System.Drawing.Printing`: Gives you classes to manage output to a print device.
- `System.Drawing.Text`: Wraps fonts and type management.

All the above given classes are available in the `System.Drawing` namespace and its associated third-level namespaces. This module actually draws UML classes. One UML class was made by combining three rectangles into a polygon. All polygons were drawn on particular points spatially located by the BPS. Various functions in VB.NET such as `Graphics.FromHwnd()`, `Rectangle()`, `DrawRectangle()` were used to draw classes. Similarly, to draw associations, generalizations, the functions like `DrawLine()`.

5 Case Study

To test the designed tool XMI2UML, we present here a solved case study. The solved case study is a sub set of a Library Domain Model (LDM) [25] that describes main classes and relationships which could be used during analysis phase to better understand domain area for Integrated Library System (ILS), also known as a Library Management System (LMS).

The problem statement consists of a UML Class model (see Figure 4) and a XMI 2.1 Representation. Both representations (UML and XMI) were generated using Enterprise Architect [24] v9.0 software. The original case study is available at [25]. However, we have solved a subset of the case study. The solved subset consists of five classes and five relationships. Fig 4 shows the subset (UML class model) of the solved case study. Following is the problem statement for the case study (See Figure 4):

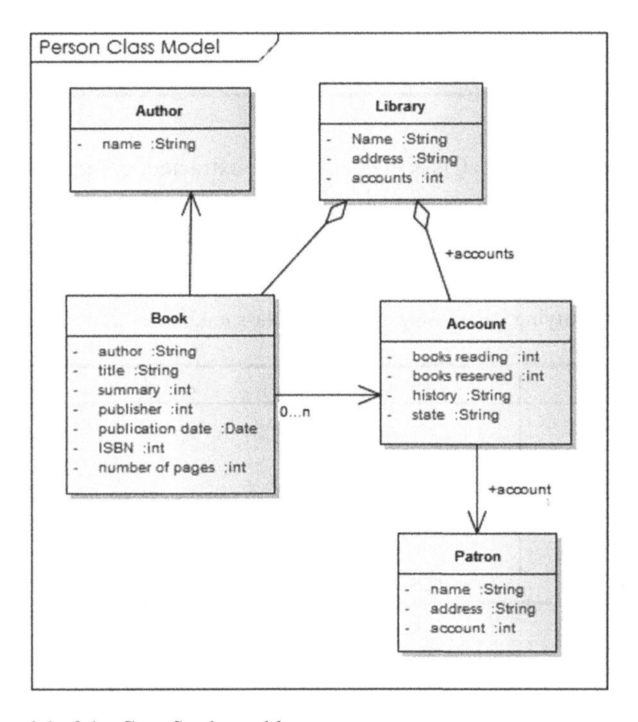

Fig. 4 UML class model of the Case Study problem statement

The problem statement of the case study was given as input (XMI representation) to the XMI2UML tool. The tool parses the input and extracts the UML class elements (see Table 1):

Table 1 Extracted UML Class Elements

Category	*Count*	*Details*
Classes	05	Author, Book, Library, Account, Patron
Attributes	18	name, author, title, summary, publisher, publication date, ISBN, number of pages, name, address, accounts, books reading, books renewed, history, state, name address account
Operations	00	-
Objects	00	-
Multiplicity	01	0..1
Associations	03	Account
Aggregations	02	Accounts
Generalizations	00	-

Once all the UML Class elements were extracted, a logical model of the target class diagram was developed. The logical model was based on relationships in all candidate classes. The generated logical model is shown in Table 2:

Table 2 Identifying relationships in a UML class model.

S# / Type	*Source*	*Mult. A*	*Label*	*Mult. B*	*Destination*
Relation 01	Book	-	-	-	Author
Relation 02	Book	-	-	-	Library
Relation 03	Book	0...n	-	-	Account
Relation 04	Account	-	Accounts	-	Library
Relation 05	Account	-	Account	-	Patron

Once the relationships in a UML class model are generated, the graphical module of XMI2UML module generates the target UML class diagram. The finally labeled UML class diagram is shown in Figure 5 where the orange doted lines are showing the binary partition of the space:

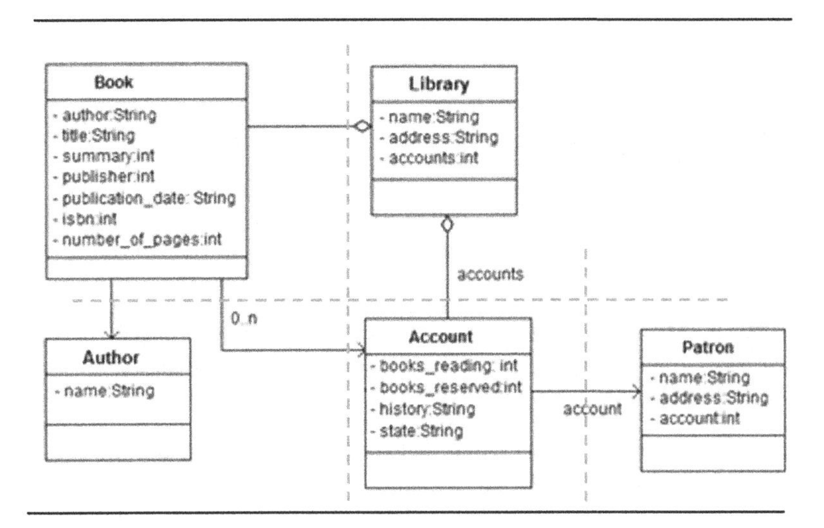

Fig. 5 Finally generated UML class model

There is an issues regarding efficiency of a BSP tree that relates to the balancing of the tree. Regarding balancing of a BSP tree, Bruce Naylor has presented some good algorithms and techniques to maintain well-balanced BSP trees. For a bigger UML class models, where we have to deal with bigger BSP trees, well-balancing of a tree become more critical.

6 Conclusions

The This research is all about designing and implementing a theory that can read, understand and analyze the XMI file and generate UML class models. The proposed system will be fully automated and able to find out the classes and objects and their attributes and operations using an artificial intelligence technique. Then the UML diagrams such as class diagrams would be drawn. The accuracy of the software is expected up to about 80% without any involvement of the software engineer provided that he has followed the pre-requisites of the software to prepare the input scenario. The given input should be an XMI file. A graphical user interface is also provided to the user for entering the Input XMI representation in a proper way and generating UML diagrams. The current version of XMI2UML tool can process XMI 2.1 version. This research was initiated with the aims that there should be software which can read the XMI representation and can draw the UML class diagrams

References

[1] OMG. XML Metadata Interchnage (XMI) version 2.1. Object Management Group (2005), http://www.omg.org
[2] OMG. Unified Modeling Language: Diagram Interchange version 2.0, OMG document ptc/03-07-03 (July 2003), http://www.omg.org
[3] Alanen, M., Lundkvist, T., Porres, I.: A Mapping Language from Models to XMI (DI) Diagrams. In: Proceedings of the 31st Euromicro Conference on Software Engineering and Advanced Applications, pp. 450–458. IEEE Computer Society, Portugal (2005)
[4] Bailey, I.: XMI, UML & MODAF (February 2005), http://www.modaf.com
[5] Dolog, P., Nejdl, W.: Using UML and XMI for Generating Adaptive Navigation Sequences in Web-Based Systems. In: Stevens, P., Whittle, J., Booch, G. (eds.) UML 2003. LNCS, vol. 2863, pp. 205–219. Springer, Heidelberg (2003)
[6] Kovse, J., Härder, T.: Generic XMI-Based UML Model Transformations. In: Bellahsène, Z., Patel, D., Rolland, C. (eds.) OOIS 2002. LNCS, vol. 2425, pp. 192–198. Springer, Heidelberg (2002)
[7] OMG: Meta Object Facility Specification (MOF) version 1.3.1, Object Management Group (2001), http://www.omg.org
[8] Wagner, A.: A pragmatically approach to rule-based transformations within UML using XMI. In: Workshop on Integration and Transformation of UML Models (2002)
[9] Jensen, M.R., Møller, T.H., Pedersen, T.B.: Converting XML Data To UML Diagrams For Conceptual Data Integration In: 1st International Workshop on Data Integration over the Web (DIWeb) at 13th Conference on Advanced Information Systems Engineering, CAISE 2001 (2001)
[10] Mich, L.: NL-OOPS: from natural language to object oriented requirements using the natural language processing system LOLITA. Natural Language Engineering 2(2), 167–181 (1996)
[11] Delisle, S., Barker, K., Biskri, I.: Object-Oriented Analysis: Getting Help from Robust Computational Linguistic Tools. In: 4th International Conference on Applications of Natural Language to Information Systems, Klagenfurt, Austria, pp. 167–172 (1998)
[12] Börstler, J.: User - Centered Requirements Engineering in RECORD - An Overview. In: Nordic Workshop on Programming Environment Research, NWPER 1996, Aalborg, Denmark, pp. 149–156 (1999)
[13] Overmyer, S.V., Rambow, O.: Conceptual Modeling through Linguistics Analysis Using LIDA. In: 23rd International Conference on Software Engineering (July 2001)
[14] Perez-Gonzalez, H.G., Kalita, J.K.: GOOAL: A Graphic Object Oriented Analysis Laboratory. In: 17th Annual ACM SIGPLAN Conference on Object-Oriented Programming, Systems, Languages, and Applications, OOPSLA 2002, NY, USA, pp. 38–39 (2002)
[15] Harmain, H.M., Gaizauskas, R.: CM-Builder: A Natural Language-Based CASE Tool for Object- Oriented Analysis. Automated Software Engineering 10(2), 157–181 (2003)
[16] Oliveira, A., Seco, N., Gomes, P.: A CBR Approach to Text to Class Diagram Translation. In: TCBR Workshop at the 8th European Conference on Case-Based Reasoning, Turkey (September 2006)

[17] Anandha Mala, G.S., Uma, G.V.: Automatic Construction of Object Oriented Design Models [UML Diagrams] from Natural Language Requirements Specification. In: Yang, Q., Webb, G. (eds.) PRICAI 2006. LNCS (LNAI), vol. 4099, pp. 1155–1159. Springer, Heidelberg (2006)
[18] Bajwa, I.S., Samad, A., Mumtaz, S.: Object Oriented Software modeling Using NLP based Knowledge Extraction. European Journal of Scientific Research 35(01), 22–33 (2009)
[19] Boger, M., Jeckle, M., Müller, S., Fransson, J.: Diagram Interchange for UML. In: Jézéquel, J.-M., Hussmann, H., Cook, S. (eds.) UML 2002. LNCS, vol. 2460, pp. 398–411. Springer, Heidelberg (2002)
[20] Ranta-Eskola, S., Olofsson, E.: Binary Space Partioning Trees and Polygon Removal in Real Time 3D Rendering. Uppsala Master's Theses in Computing Science (2001)
[21] Fuchs, H., et al.: Near Real-Time Shaded Display of Rigid Objects. Computer Graphics 17(3), 65–69
[22] Ize, T., Wald, I., Parker, S.: Ray Tracing with the BSP Tree. In: IEEE Symposium on Interactive Ray Tracing, RT 2008, pp. 159–166 (2008)
[23] Chen, S., Gordon, D.: Front-to-Back Display of BSP Trees. IEEE Computer Graphics & Algorithms, 79–85 (September 1991)
[24] Matuschek, M.: UML - Getting Started: Sparx Enterprise Architect, To a running UML-Model in just a few steps, by Willert Software Tools (2006), `http://www.willert.de/assets/Datenblaetter/UML_GettingStarted_EA_v1.0en.pdf`
[25] Library Domain Model (LDM), `http://www.uml-diagrams.org/class-diagrams-examples.html`
[26] Bajwa, I.S., Amin, R., Naeem, M.A., Choudhary, M.A.: Speech Language Processing Interface for Object-Oriented Application Design using a Rule-based Framework. In: International Conference on Computer Applications, ICCA 2006, pp. 323–329 (2006)
[27] Bajwa, I.S., Choudhary, M.S.: Natural Language Processing based Automated System for UML Diagrams Generation. In: Saudi 18th National Conference on Computer Application, NCCA 2006, pp. 171–176 (2006)
[28] Bajwa, I.S., Naeem, M.A., Ali, A., Ali, S.: A Controlled Natural Language Interface to Class Models. In: 13th International Conference on Enterprise Information Systems, ICEIS 2011, Beijing, China, pp. 102–110 (2011)

Distributed Store for Ontology Data Management

Abdullah Alamri and Peter Bertok

Abstract. The Semantic Web is recognized as the next generation web which aims at the automation, integration and reuse of data across different internet applications. To better understand and utilize the Semantic Web, the W3C adopted standards and tools such as the Resource Description Framework (RDF) and Web Ontology Language (OWL). Management of large amounts of semantic data, stored in semantic models, are required within almost every semantic web application environment, thus motivating the design of specific repositories in order to store and operate semantic models. However, most semantic storage systems based on a relational database support both a monolithic schema with a single table that stores all statements. Since the Semantic Web systems use OWL ontologies as set of RDF triples (which are not composed of hierarchical knowledge between class and property) rather than complex OWL data models, they are not compliant with the more advanced systems. Further, an expected reduced system performance can be observed due to the large amount of semantic data managed in a single storage model. We propose improving the storage model of OWL by changing the management of OWL data and designing an efficient new relational database layout to store an OWL knowledge base, focused on the OWL 2 DL language relational. Upon evaluation, our storage system shows an improvement in the performance and management.

Keywords: Semantic Web, RDF, OWL and Semantic Repositories.

1 Introduction

A growing number of domains are adopting semantic models, as a centralized gateway, in order to achieve semantic interoperability among data sources and

Abdullah Alamri · Peter Bertok
School of Computer Science and Information Technology, RMIT University, Melbourne, Australia
e-mail: abdullah.alamri@rmit.edu.au

R. Lee (Ed.): Computer and Information Science 2012, SCI 429, pp. 15–35.
springerlink.com

applications, or directly for modelling and managing relevant information. The concept of the Semantic Web refers to the World Wide Web Consortium's (W3C's) vision of the web of data that enables individuals to construct data stores on the web, build vocabularies and write rules for handling data; so that it can be understood and used by machines for automation, integration, and the re-use of data across various applications [5]. The purpose of the Semantic Web, envisioned by Berners-Lee, is not only to connect the seemingly endless amount of data on the World Wide Web, but also to enable connections for data information, within databases and other non-interoperable information repositories [5]. The main technologies that make up the Semantic Web are the Resource Description Framework (RDF) and the Web Ontology Language (OWL), which are currently utilized to store data and information [17]. OWL ontology became a W3C recommendation mechanism for creating Semantic Web applications and is considered as the most promising semantic ontology language built upon the syntax of RDF and RDF schema; it also facilitates the sharing of information by identifying the types of relationships that can be expressed by using RDF/XML syntax to explain the hierarchies and relationships between different resources and can be utilized as a data repository [17].

Most of the semantic web applications developed for such an environment require the management of large amounts of semantic data, which are stored in semantic models; therefore, the efficient management of the semantic models is a critical factor in determining the performance of many applications. This requirement has served to motivate the design of specialized repositories, in order to store and manipulate semantic models. A semantic repository is a system that has at least two features: (i) it offers a triple store facility for RDF data sources, since ontologies are often expressed in RDF; (ii) it provides an infrastructure for the semantic inference engine to derive answers from the data source for both explicit and implicitly-derived information.

The problem investigated in this paper is that most OWL storage systems based on a relational database support a monolithic schema with a single table that stores all statements. Furthermore, these systems persist OWL ontologies as sets of RDF triples and do not include hierarchical knowledge among classes and properties and hence they are not compliant with the complex data model of OWL expressions [2,9,12,18]. This also indicates some issues, for example, an expected reduced system performance can be observed due to the large amount of data managed in a single storage model (i.e. single table). This paper addresses the research question of **how OWL data can be efficiently managed and stored in a persistent semantic storage to improve the performance of semantic repositories**.

The data representation in triple store, though flexible, has the potential for serious performance issues, since there is only a single table that store all statements. In this paper, we developed an efficient relational database layout to store an OWL ontology knowledge base. Hence, it is our goal in this paper to explore ways to improve the query performance, so the **contributions** of our research work documented in this paper are:

- Distributed store based on Tbox, Abox and OWL constructors: Semantic information is kept in various places for reasons of system design and performant implementation.
- Designing an efficient relational database model to store OWL data using clustering table technique which provides a method for clustering semantic data into different categories as follows:
  - Primitive OWL entities clustering. It contains the declaration of domain terminology and relations in the form of OWL entities (owl:Class; owl:ObjectProperty; owl:DatatypeProperty).
  - ABox Clustering. The ABox of an ontology contains all statements about class instances and values. The ABox-Cluster consists ground sentences stating where in the hierarchy individuals belong (i.e., relations between individuals and concepts).
  - TBox Clustering. The TBox consist the Schema-Level. The Schema-Level contains all custom class and property definitions. It furthermore defines their relations, describes their meanings and is therefore the basic source of information about resources.
  - OWL's complex class constructor clustering. It contains disjunction, negation and value restrictions, as well as numerous large and complex full-class definitions.
- The OWL storage model is proposed with well-defined rules to address the drawbacks of earlier ontology storage techniques. While there are many solutions and tools for persistent ontologies in databases, most of them are typical representatives of triple based storage systems; while other methodologies are based on some kind of (partial) ontology meta-model.
- The proposed storage system scheme is based on OWL 2 DL that represents all constructs of an ontology document.
- We conduct experiment in order to evaluate the proposed storage model by studying the query performance of the proposed structure. Furthermore, we compare the querying cost and the query performances of the proposed model. The results show that OWL database schema for high performance inferencing is by means of the derived queries.

The rest of the paper is organized as follows: in section 2, we will cover basic information about related paradigms and technologies. Section 3 provides a description of the overall model structure. Section 4 & 5 presents the implementation, performance analysis and experimental results. Section 6 concludes the paper.

2 Background

This section describes the main concepts related to the work presented in this paper.

2.1 Survey of the Web Ontology Model

Ontologies are metadata schemas that provide a controlled vocabulary of concepts and allow the encoding of knowledge about specific domains [14]. They often include reasoning rules that support the processing of that knowledge. A number of ontology languages have been developed for the representation of ontologies on the World Wide Web. Figure 1 shows a brief history of web ontology models. Some ontology languages include simple HTML ontology extensions (SHOE), the Resource Description Framework (RDF), the ontology interchange language (OIL), the DARPA Agent Mark-up Language (DAML) and Web Ontology Language (OWL) [4, 16]. The Web Ontology Language (OWL) has become a W3C recommendation and standard ontology language for creating semantic web applications. OWL is a revision of the DAML+OIL web ontology language with various levels of expressivity included in the language. There are three species of OWL ontologies: **OWL Lite, OWL DL,** and **OWL Full**. Every OWL Lite ontology is also an OWL DL ontology, and every OWL DL ontology is a legal OWL Full ontology. in this paper, we focus on the OWL 2 language.

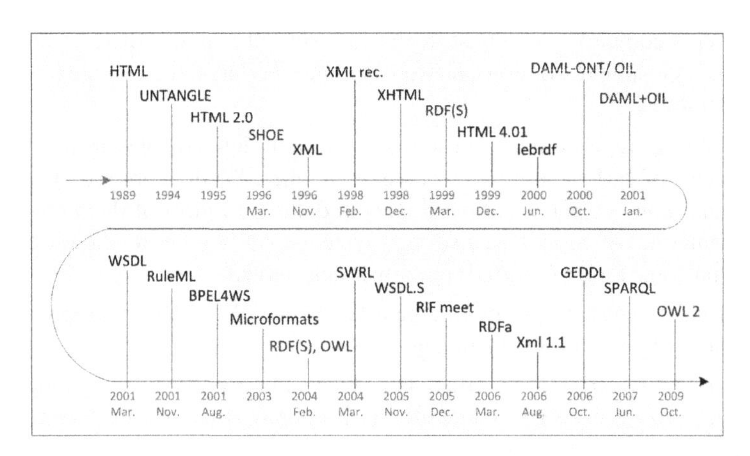

Fig. 1 A number of the ontology languages that have been developed for the representation of ontologies on the World Wide Web

2.2 OWL 2

OWL 2 is a recent extension of the Web Ontology Language and is compatible with the original OWL standard now referred to as "OWL 1". The W3C Consortium recommended OWL 2 as a standard for ontology representation on the Web on October 27, 2009. All varieties of OWL use RDF for their syntax, so, as in OWL 1, the main syntactic form of an OWL 2 ontology is based on RDF serialization and also some alternative syntactic forms are included in OWL 2. OWL 2 elements are identified by Internationalized Resource Identifiers (IRIs). It extends OWL 1 which uses Uniform Resource Identifiers (URIs). There are new features in OWL 2: extra syntactic sugar for making some statements simple, extended data type support, simple meta-modelling, additional property and qualified cardinality constructors as well as extended annotations are among the new features of OWL 2 [13]. OWL 2 stipulates profiles that are tractable, and these include OWL 2 DL, OWL 2 QL, OWL 2 EL and OWL 2 RL. These profiles are all based on description logics (DLs) [8, 14, 19]. In light of the considerable progress that has been achieved in OWL 2, this paper analyse those concepts of OWL 2 DL. OWL 2 DL is the most expressive profile in OWL 2 which is based on the description logic SROIQ, and is geared towards enabling ontologies with a high degree of expressivity in the language. It has a fair amount of extra features in comparison to OWL 1 DL for the purpose of retaining decidability in exchange for more modelling features. It can be regarded as the full expressivity of OWL 2 obtained under OWL 2 direct semantics.

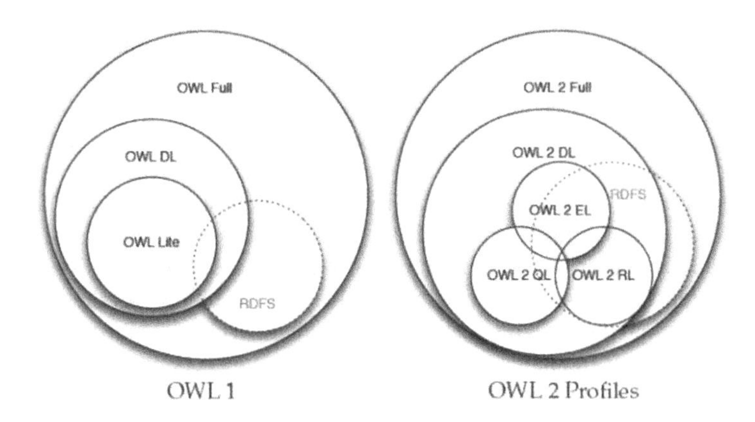

Fig. 2 OWL 1 and OWL 2 Syntactic Subsets (Profiles)

2.3 Semantic Repositories

Along with popular research on the semantic web, semantic repositories have been developed using several different approaches to store RDF/OWL data. In general, three main storage types can be identified, based on the persistent strategy they use: in-memory storage, file system storage and relational database [2, 18]. Especially, most semantic storage systems based on a relational database, because it is widely used for data management and is stable through improvement over several decades [9, 12].

Currently, there are several semantic repositories available as both open-source projects and proprietary solutions such as Sesame [7], Jena [18], Kowari Metastore [21], AllegroGraph [20] and Virtuoso [18]. Jena provides a programmatic environment for RDF or OWL and offers methods to load RDF data into a memory-based triple store, a native storage or into a persistent triple store. In brief, from RDF(S) to OWL, Jena provides a storage model based on a relational database that has a single storage model (stores most data in a table), which cannot satisfy the complex data model of OWL data and causes many problems. For instance, a large amount of redundant data will be produced, and this will also affect system performance [2, 18]. Sesame is an open source RDF database that provides storage and querying facilities for RDF Schema and OWL [7]. Sesame Supported features include several query languages (SeRQL and SPARQL), RDF Schema inference engine and memory, native disk or relational database storage. In the case of relational database storage, a permanent OWL ontology storage system using relational databases supports both a monolithic schema with a single table that stores all statements. However, these systems still have problems related to hierarchical relations as OWL data cannot be directly represented by a single relational table [7]. The Virtuoso Universal Server is a middleware and database engine hybrid that provides SQL, XML, and RDF data management in a single, multithreaded server process. Virtuoso provides persistent storage to manage an OWL knowledge base in relational databases. According to the evaluation in [18], Virtuoso still suffers from insufficiency. Moreover, there are various RDF storage systems implemented with different methods and targets such as ARC, Redland, and RDFstore. However, they persist OWL ontologies as a set of RDF triples absent of hierarchical knowledge between class or property and do not consider specific processing for complex class descriptions generated by class constructors [18]. Furthermore, many attempts have already been taken to map ontologies into relational databases [1,11,15]. These previous work suggested different approaches to deal with mapping between ontologies and relational databases or transformation. However, these approaches suffer from some limitations. For instance, the mapping rules did not refer to all the OWL elements (they ignored the design mapping rule of unnamed class "value restrictions") or stored an ontology and its instances in the same manner (one fact table).

3 Proposed Method

In this section, we provide a description of the overall model structure. We formalized the data model of OWL 2 DL, followed by a description of the whole structure of the proposed storage model.

3.1 Data Model of OWL 2 DL

In order to design an efficient relational data model to store OWL 2 DL data persistently, we briefly look at the OWL 2 DL data model and define an OWL 2 DL ontology as follows:

Definition 1: An ontology is a structure $\mathcal{O} = (\mathcal{C}, \mathcal{OP}, \mathcal{DP}, \mathcal{I}, \mathcal{E}, \mathcal{T}, \mathcal{A})$, where

- $\mathcal{C}$ is a set of concepts in the ontology (classes in OWL 2 DL).
- $\mathcal{OP}$ is a set of object properties as defined in the OWL 2 DL Specification.
- $\mathcal{DP}$ is a set of data properties as defined in the OWL 2 DL Specification.
- $\mathcal{I}$ is a set of instances (also called individuals of the concepts).

- $\mathcal{E}$ is a set of expressive class constructors, these constructors can be used to create so-called class expressions.
- $\mathcal{T}$ is a set of TBox axioms that provide information about classes and properties.
- $\mathcal{A}$ is a set of ABox facts that represent the assertions (or statements) in the domain knowledge base.

Class Elements ($\mathcal{C}$)

OWL provides mechanisms to represent all the components of ontology: classes, properties (or relations), instances, class constructors, axioms and facts [8, 17]. Classes represent an abstraction mechanism for grouping instances with similar characteristics (properties). Classes in OWL are usually organized in a specialization hierarchy (taxonomy) based on the superclass-subclass relation. OWL comes with two predefined classes:"`owl:Thing`" which is the most general class $\top$, the root class and "`owl:Nothing`" is an empty class $\bot$ that has no instances. "Consequently, every OWL class is a subclass of owl:Thing, and owl:Nothing is a subclass of every class" [3].

Property Elements ($\mathcal{OP}$ & $\mathcal{DP}$)

In OWL, a property is a binary relation connecting concepts and can be further distinguished as *Object Property* or *Data Property*.

- *Object property* represents the relation between instances of two classes (relate instances to other instances),
- *Data property* represents the relation between instances of classes and literal values such as xsd:number, xsd:string, and xsd:date.

Properties can have a specified domain and range. OWL uses built-in datatypes supported by XML Schema, which is referenced using: *http://www.w3.org/2001/XML-Schema#name*. The following example shows the *data property* specifying Course title is xsd:string.

```
<DataPropertyDomain>
  <DataProperty IRI="Title"/>
  <Class IRI="Course"/>
 </DataPropertyDomain>
 <DataPropertyRange>
  <DataProperty IRI="Title"/>
  <Datatype IRI="http://www.w3...#String"/>
 </DataPropertyRange>
```

OWL also allows the specification of property characteristics such as Transitive, Symmetric, Asymmetric, Functional, Inverse Functional, Reflexive, Irreflexive. More details in Ref. [4, 16].

Instances ($\mathcal{I}$)

Instances represent individuals in an ontology and are defined as the actual entities. The following example shows Jason is an instance of postgraduate student concept.

```
<ClassAssertion>
    <Class IRI="PostgraduateStudent"/>
    <NamedIndividual IRI="Jason"/>
</ClassAssertion>
```

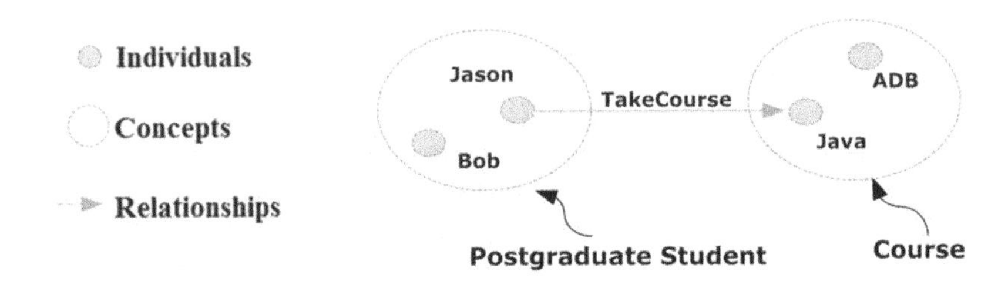

Fig. 3 An example of OWL Ontology which contains classes and instances, relations between instances & classes

Class Expressions ($\mathcal{E}$)

OWL supports various ways to describe classes: the intersection or union of two or more class descriptions, the complement of a class description, the enumeration of individuals that form class instances and property restrictions. OWL defines two kinds of property restrictions: *value constraints* and *cardinality constraints*, which are special kinds of class description. A value constraint (e.g., allValuesFrom, someValuesFrom and HasValue) puts constraints on the range of the property when applied to this particular class description. A cardinality constraint (e.g., maxCardinality, minCardinality and cardinality) is commonly used to constrain the number of values a property can take, in the context of this particular class description.

OWL TBox Axioms ($\mathcal{T}$)[1]

Definition 2: Let T be TBox is a set of terminological axioms describing class and property hierarchies.

- Class expression axioms allow the establishment of relationships between class expressions.
- Object property axioms allow relationships to be characterized and established between object property expressions
- Data property axioms establish relations between data properties.

[1] OWL 2 axioms and facts (Source: Ref. [13])

ABox Facts (Assertions) ($\mathcal{A}$)

Definition 3: ABox A is consistent with TBox T which contains a finite set of assertions about individuals in the form of classes that the individual belongs to or its relationship to other individuals or about individual identity.

3.2 OWL Data Storage

In this section, we define a database layout to efficiently store an OWL ontology knowledge base and describe the whole structure of the proposed storage model. Our relational database schema proposed in this paper can be defined as Definition 4.

Definition 4: An OWL knowledge base O is managed as four categories based on the data model as follows:

- Primitive OWL entities clustering. It contains the declaration of domain terminology and relations in the form of OWL entities (owl:Class; owl:ObjectProperty; owl:DatatypeProperty).
- OWL's complex class constructor clustering. It contains disjunction, negation and value restrictions, as well as numerous large and complex full-class definitions.
- TBox Clustering. The TBox consist the Schema-Level. The Schema-Level contains all custom class and property definitions. It furthermore defines their relations, describes their meanings and is therefore the basic source of information about resources.
- ABox Clustering. The ABox of an ontology contains all statements about class instances and values. The ABox Clustering consists ground sentences stating where in the hierarchy individuals belong (i.e., relations between individuals and concepts).

Many semantic repositories use one table to layout their storage system. The trick is to break up the table into layers. Hence, the proposed storage model manages different types of triples using different tables. The tables of the database schema have been categorized into four semantic clustering (layers) As illustrated in Figure 4. This categorization is done based on the fundamental OWL structures: classes, object properties, data properties and individuals, as well as OWL class constructors in order to express more complex knowledge, and OWL axioms that give information about classes and properties and facts which represent individuals in the ontology, the classes they belong to, the properties they participate in and individual identity.

The proposed process rules for mapping an ontology to a relational database is summarized in Algorithm 1.

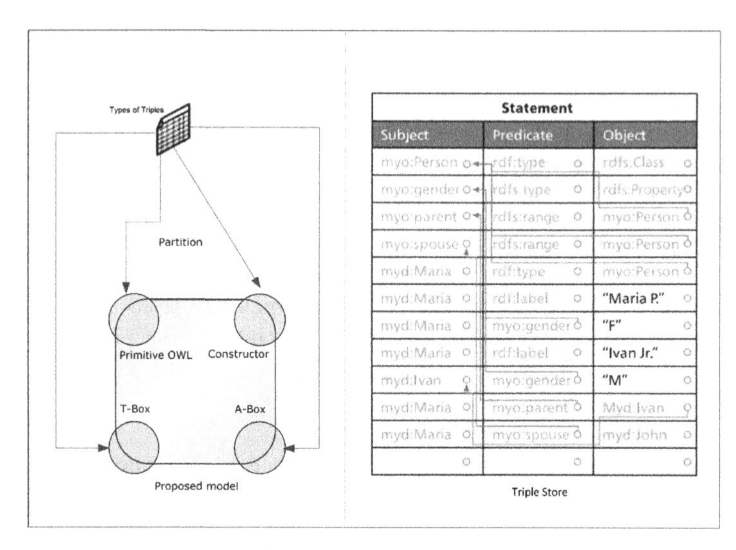

Fig. 4 The Proposed Method

Algorithm 1. OWL_RDMS()

1: **Data**: OWL //OWL ontology
2: **Result**: OWLRD //OWL ontology relational database 'Definition 2'
3: Start Algorithm
4: Reading OWL document;
5: Check for consistency and completeness by analyzing the structure of OWL data;
6: Extracting OWL data;
7: Classifying the data according to the meanings of OWL data;
 - Classifier OWLDataClassify(OWL Data)
8: Create mapping objects with auto mapper;
 - OWLDataClassified ← CreateMappingObject(Mapper)
9: **Execute** the mapping rules to load OWL data into OWL relational database as described below;
10: The final output, OWLRD

A PRIMITIVE OF OWL ENTITIES: This layer contains the declaration of domain terminology and relations in the form of OWL entities (owl:Class; owl:ObjectProperty; owl:DatatypeProperty). Classes, object properties, and data properties are stored in class, object property, data property tables, respectively. A class table is composed of ClassID that is generated to be unique across all ontologies and ClassName indicates the name of the class. The object property table contains ObjectPropertyID, PropertyName and its characteristics that allow the meaning of properties to be enriched through the use of property characteristics: (inverse) functionality, transitivity, symmetry, asymmetry, reflexivity, irreflexivity property.

Data property table contains DataPropertyID, PropertyName and its characteristics (FunctionalDataProperty).

OWL'S COMPLEX CLASS CONSTRUCTORS: This layer contains a number of operators for constructing class descriptions out of logical statements. There are three types of class descriptions: Boolean(Union, Intersection and Complement). Additionally, classes can be enumerated (owl:oneOf) and property restrictions. These types of class descriptions result in the definition of a new class that is a subclass of owl:Class and is not necessarily a named class (i.e. 'anonymous' class) as defined in the OWL syntax. So, these tables are defined to deal with OWL anonymous classes that arise in various situations. We designed mapping rules for this complex class constructors as follows:

If class C_i is defined by intersectionOf, then we generate a rule to create a table for every intersection class where the primary key is ID of C_i, and other columns are elements of intersectionOf in turn. There is a similar rule for union classes. Because intersectionOf and unionOf are not always binary operations, we create a table for every complex class.

If class C_i is defined by complementOf, we generate a rule for all the complementOf relations whose primary key is ID of C_i, other column define the complement of the class. Because complementOf relations is always a binary operation, we create a table for all complementOf relations.

If a class C_i is an enumeration, then we generate a rule to create one table for each enumerated class because oneOf is not always a binary operation, where the primary key is ID of C_i, and other columns are an enumeration of named individuals.

For all the restrictions in OWL, we generate a rule to create a table for restriction classes where ObjectAllValuesFrom, ObjectSomeValuesFrom and ObjectHasValue restrictions have their own metadata table. These tables contain a new class ID designed as the primary key of the table and OnPropertyID (objectProperty) which links to the object property table. ObjectAllValuesFrom$_{table}$, ObjectSomeValuesFrom$_{table}$ has column restrictionClass ID which points to the table of the corresponding restriction source class. ObjectHasValue$_{table}$ has the column individual ID. ObjectCardinality also has their own metadata table which contains CardinalityClass ID designed as the primary key of the table, OnPropertyID on which the restriction is applied and minCardinality, maxCardinality and Cardinality as one column of the table in turn. There is a similar rule for data cardinality restriction. In regard to data property value restriction, DataAllValuesFrom, DataSomeValuesFrom and DataHasValue have their own metadata table where a new class ID is designed as the primary key of the table and OnDataPropertyID which links to the data property table. DataAllValuesFrom$_{table}$, DataSomeValuesFrom$_{table}$ has column restricted data range and DataHasValue$_{table}$ has the column "value" for storing the literal value.

TBOX: This layer contains axioms which are used to associate class and property IDs with either partial or complete specifications of their characteristics, and to give other logical information about classes and properties. Class Axioms: There are three types of class axioms: Subclass, Equivalent class and Disjointness class. If there is a subclass relationship between two concepts, then we generate a rule to map all the class inheritance to SubClass$_{table}$ which has two columns, one is called C_i-subclass-, the other is C_j. Also, if two classes are equivalent or disjoint, then their relation is done in EquivalentClass$_{table}$ (C_i-equivalent-, C_j) , or DisjointClass$_{table}$ (C_i-Disjoint-, C_j). Note all the concept IDs are assigned as foreign keys. (C_i, C_j) is designed as the primary key of the table. Also, there is DisjointUnion axiom which allows to define a class as the union of other pairwise disjoint classes. For example, the class Parent is the disjoint union of the classes Mother and Father. Information about groups of disjoint union classes is saved in tables DisjointUnion where DisjointUnID is designed as a primary key of the table, $\mathcal{C}$_ID is conceptID assigned as foreign keys and $\mathcal{C}_i$disjointUnionID is the disjoint union of the classes. Object Property Axioms: If the object property is SubObjectPropertyOf, EquivalentObjectProperty, DisjointObjectProperty or InverseObjectPropertyOf some other object property, then the entry of their relation is done in SubObjectProperty$_{table}$, EquivalentObjectProperty$_{table}$, DisjointObjectProperty$_{table}$ or InverseObjectProperty$_{table}$. For example, if object property $\mathcal{OP}_i$ is a subObjectproperty of $\mathcal{OP}_j$, we generate a rule to map the object property inheritance into the table called SubObjectProperty$_{table}$, which has two columns, one is called $\mathcal{OP}_i$-subobjectproperty-, the other is $\mathcal{OP}_j$ which are assigned as a foreign key. ($\mathcal{OP}_i$, $\mathcal{OP}_j$) is designed as the primary key of the table. Furthermore, the OWL construct ObjectPropertyChain in a SubObjectPropertyOf axiom allows a property to be defined as the composition of several properties. More precisely, SubObjectPropertyOf(ObjectPropertyChain($\mathcal{OP}_1$ ... $\mathcal{OP}_n$) $\mathcal{OP}$) states that if an individual x is connected with an individual y by a sequence of object property expressions $\mathcal{OP}_1$, ..., $\mathcal{OP}_n$, then x is also connected with y by the object property expression $\mathcal{OP}$. For example, one of the most common examples of a property that is defined in terms of a property chain is the hasUncle relationship SubPropertyOf(ObjectPropertyChain(:hasFather :hasBrother) :hasUncle) states that if person 1 hasFather person 2 and person 2 hasBrother person 3, then person 1 hasUncle person 3. Hence, ObjectPropertyChain$_{table}$ represents the connection between the sequence number of some component property in the property chain. Furthermore, in regards to the object property domain and range information, it is mapped to ObjectPropertyDomain$_{table}$ and ObjectPropertyRange$_{table}$. In object property domain and range relation, $\mathcal{OP}$ID and $\mathcal{C}_i$ID are assigned as foreign keys. Mapping rules of data property axioms are similar to object property axioms. For example, if data property $\mathcal{DP}_i$ is a subdataproperty of $\mathcal{DP}_j$, we generate a rule to map the data property inheritance into the table called SubDataProperty$_{table}$, which has two columns, one is called $\mathcal{DP}_i$-subdataproperty, the other is $\mathcal{DP}_j$. ($\mathcal{DP}_i$, $\mathcal{DP}_j$) is designed as the foreign key. A similar rule is applied to equivalent and disjoint data property axioms.

ABox: This layer contains all statements about class instances and values. To give a clear illustration, individual is an instance of a particular class, for example John is an instance of the person class. Every individual in the ontology with the class it belongs to is mapped to Class Assertion$_{table}$ which has three columns: IndividualID, IndividualName and ClassID. Moreover, an individual can have sameAs and differentFrom relationship with other individuals. For example, if two individuals are same then their relationship is stored in the sameIndividual table. There is a similar rule for differentFrom relation. Also, individuals can have an object property relationship with other individuals, then their relationship is stored in an object property assertion table with individualID1, ObjectPropertyID and IndividualID2. Also, individuals can be connected by a data property relationship to the literal value, then their relationship is stored in a data property assertion table with individualID, DataPropertyID and DataTypeValue.

Key Benefits and Reasons to store the TBox - ABox Split
So, to conclude this part, here are some of the key benefits to store instances (the ABox) and (the TBox) that describes the structural and the component of conceptual relationships separately:

- The improved semantic storage system is efficient since some self-joins on a big triple table are changed to some joins among small-sized tables.
- We are able to handle ABox semantic data simply. The nature of instance "ABox" is comparatively constant and can be captured with easily understandable attribute-value pairs.
- ABox instance data evaluations can be done separately from conceptual evaluations, which can help through triangulation in such tasks as disambiguation or entity identification.
- Ontologies (focused on the TBox) are kept simpler and easier to understand.
- Semantic information is kept in various places for efficiently represent, reason and query semantic data Abox (instance data) in relational databases for very expressive description logic.

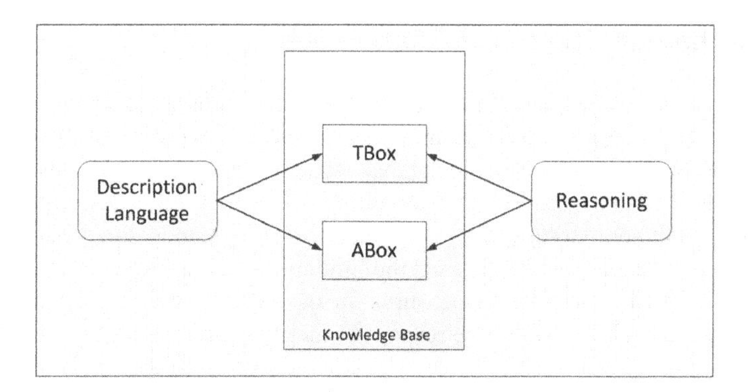

Fig. 5 The TBox - ABox Split

4 Implementation

In order to demonstrate our ontology storage method proposed in this paper, we have designed and built a prototype for our approach. All the rules for all ontology components of ontology have been implemented. The prototype presented transmits an ontology and stores it in a relational database according to the representation we have proposed. We have implemented our storage prototype based on the OWL API. MySQL was selected as a relational database, and we use it in this prototype to store the ontology knowledge base. The reasoning part is implemented in SQL and uses recursive SQL statements. At first, the ontology document is sent to the OWL validator that can be effectively used to verify the ontology syntax. After verifying the ontology syntax, the OWL document is sent to the OWL parser which can parse class expressions, class, property and individual descriptions and complete ontologies. The OWL Classifier classifies data according to the meaning and semantic relationships working together with the mapper object which was developed to map each element of OWL data to the corresponding table and then store it in the relational database by converting it into the form to store in the relational database designed in this paper. This transformation is done by applying the set of rules proposed in this paper. The following subsections summarize our experience with supporting ontology-related functionality using MySQL RDBMS.

Software and Hardware: The benchmarking was performed on a Dell inspiron machine with Intel(R) Core 2 Due CPUs (2.93GHz) with 4GB of RAM and 500GB hard drive. The stores were run under a Windows 7 operating system and using Java 1.6 with 1GB of maximum heap size and an underlying MySQL Server (version 5.1) running on the same machine. In order to analyze and evaluate the performance of the triple store storage model, we also have implemented the triple store Jena[2]-SDB version 1.3.2[3] that uses relational databases ("MySQL") for the storage of RDF and OWL, and Joseki[4] which is a query server for Jena.

5 Experimental Results and Discussion

In this section, we show querying capabilities' experimental research when ontologies are stored in the proposed semantic storage model according to the proposed mapping approach. We analysed the database schema, metadata tables, built the ontology model, and executed SQL for obtaining the results. As a proof-of-concept, the experimental methodology aims to evaluate the storage model in order to investigate and point out the capabilities and the limitations of our proposed approach.

In order to obtain comparable measures for the storage system, we made a collection of queries and measured the response time. We selected the Wine, UnONB,

---

[2] http://www.openjena.org
[3] http://www.openjena.org/SDB/
[4] http://www.joseki.org

Table 1 Benchmark ontologies' series

Ontology	Classes	Properties	Individuals
Wine	138	17	207
University (UnONB)	34	29	811
AEO	260	47	16
Family	18	17	12

AEO, Family ontologies to test on ontologies of different complexity. The statistical facts of those ontologies are shown in Table 1.

Wine Ontology is the example provided by W3C's document [17]. Due to the comprehensive and balanced utilization of diverse expressions of OWL, the Wine Ontology meets the requirements as an indicator regarding the OWL support's grade. It is utilized in determining and explaining the use of the various constructors of OWL. In addition, it consists of a structure of quite high complexity that makes heavy use of constructors like cardinality restrictions and disjunctions.

University Ontology defines fundamentals regarding the descriptions of universities and the activities that take place at such universities. The specification of the ontology gives basic concepts and properties for describing University concepts and covers a complete set of OWL constructs, respectively.

The Athletics Event Ontology (AEO) is an ontology which explains athletic events of track and field. It was generated within BOEMIE, an EU-funded project. The ontology's conceptualization is founded on the guide of the International Association of Athletics Federations (IAAF), which explains the athletic events' rules as well as regulations. The ontology is somewhat large and consists of a high DL expressivity [10].

The family ontology is an example to test the new features in OWL 2. This ontology, though, does not utilize all of OWL 2's innovative aspects. In order to test all the new features, we therefore make the addition of more ontology statements [13]. These ontologies and instance data sets were utilized within the evaluation. We performed all tests utilizing the same RDBMS. In order to evaluate the performance of the storage model, we carried out the testing using the different complex queries which require data from different data sources. We also measured completeness and correctness of the query results to decide if a repository is correct, complete or both.

- Statistical Queries: This category contains queries to get all classes and search all hierarchical instances of a specific class.
- Assertion Queries: This category consists of queries for the assertions of a given individual.
- Axiom Queries: We also queried for particular OWL axiom types, such as subclasses, object properties domain and range, disjoint Classes.

The following metrics were used to test the proposed model on different data set complexity:

- Response time. The time to issue a query, obtain results, and iterate through them.
- The correctness of the results. In practice, this was checked simply by comparing the actual and expected number of results.
- Whether a run resulted in a timeout or error.

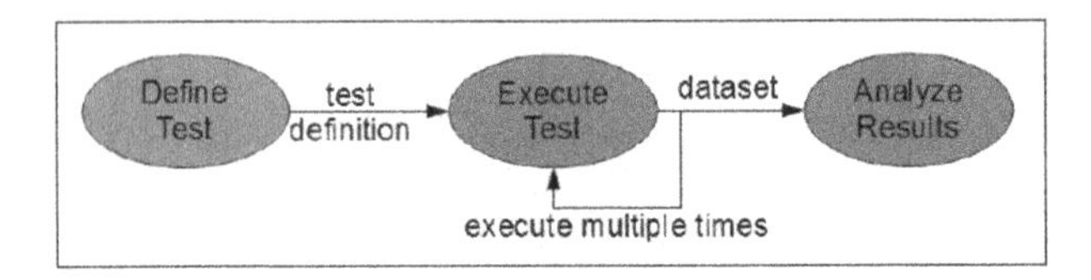

Fig. 6 The overall processing flow, which includes the 3 steps involved in running a test

Query response time is measured based on the process used in database benchmarks. To account for caching, each query is executed three times consecutively and the average time is computed. The figures 7, 8 and 9 below illustrate the execution time of the proposed model which demonstrates reasonable performance even with the increased number of data set. Furthermore, our analysis of the query results established that the proposed model was sound for the dataset queries, as the query results were correct for the corresponding queries.

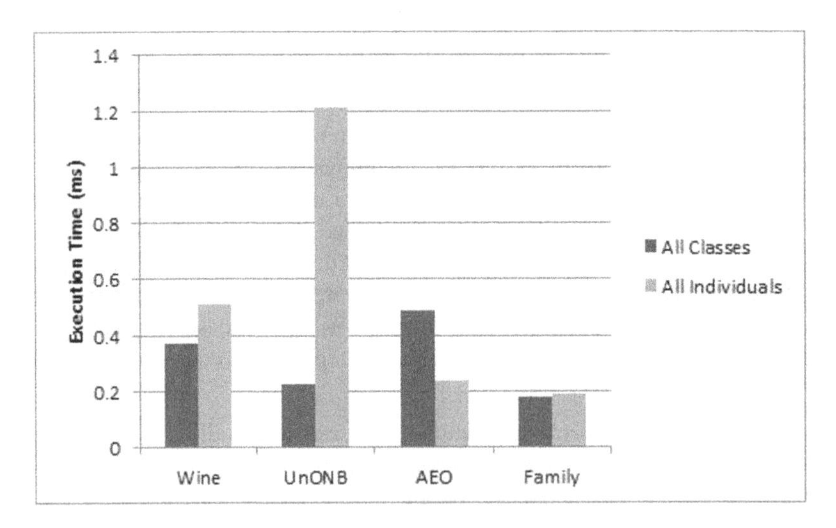

Fig. 7 Execution time on statistical queries for various ontologies

Comparison with Triple Store: As a proof-of-concept, test queries were executed against a semantic storage system (Jena 'MySQL') in order to investigate and point out the capabilities and the shortcomings in comparison with the storage model

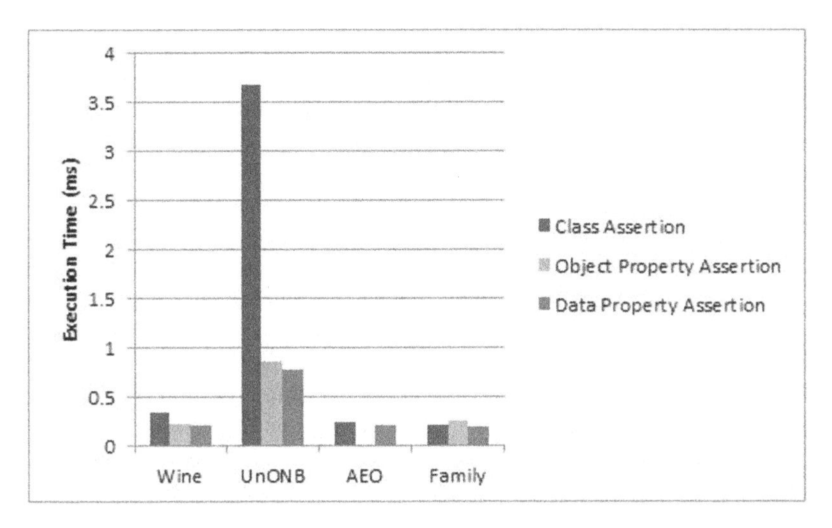

Fig. 8 Execution time on assertion queries for various ontologies

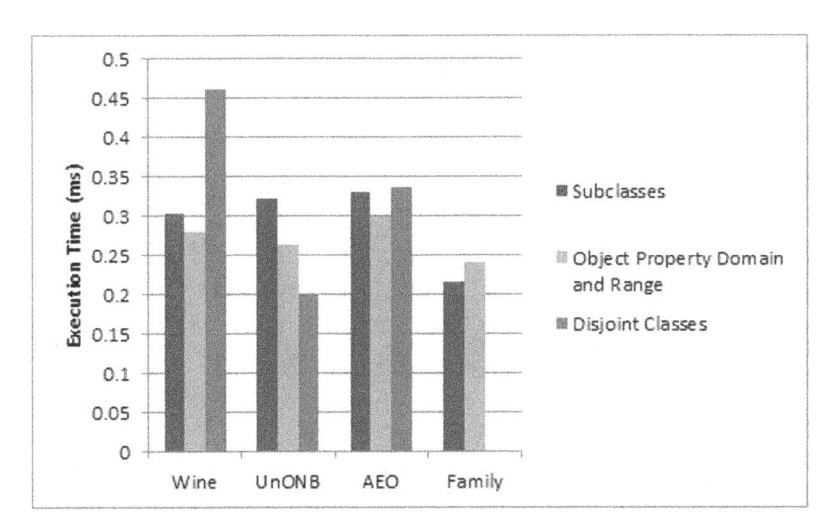

Fig. 9 Execution time on axiom queries for various ontologies

designed in this paper. The BSBM benchmark framework was chosen to provide information about the performance test [6]. The test was performed on a standard benchmark dataset ("an e-commerce domain in which a set of products is offered by different vendors and consumers have posted reviews about these products on various review sites") to examine the ontology storage with characteristic : a large number of triple patterns (one million explicit statements) and with university dataset to test other complex relations with a range of 12885 and 21725 triples. The impact of the optimization of the query plan, layout of the storage, as well as caching, as it pertains to the overall query performance, relies highly on the system's concrete configuration and on the kinds and number of queries which supplies in filling the caches. Thus, in order to have the capability to report meaningful benchmark outcomes, we consequently optimized the system's configuration as well as warmed up the system's caches through the execution of query mixes in anticipation of the average runtime for each query mix that was stabilized. The system's query performance was calculated through the running of 50 warm up mixes using one client against the system's storage. Additionally, the storage system ran on the same machine in an effort to reduce the network latency's influence. In order to obtain a bird's eye view pertaining to the performance of the stores, Figure 10 below provides query evaluation results against the 1M BSBM dataset versions. Figure 11 depicts query performance results for the university dataset. Figure 12 depicts the response times, averaged, and were summed over the entire amount of queries as well as plotted against the total number of triples. The average response times were computed over the three replicates and summed over all the queries. Consequently, the results demonstrate that the storage model designed in this paper outperforms most queries and demonstrates reasonable performance for constructing a semantic storage model.

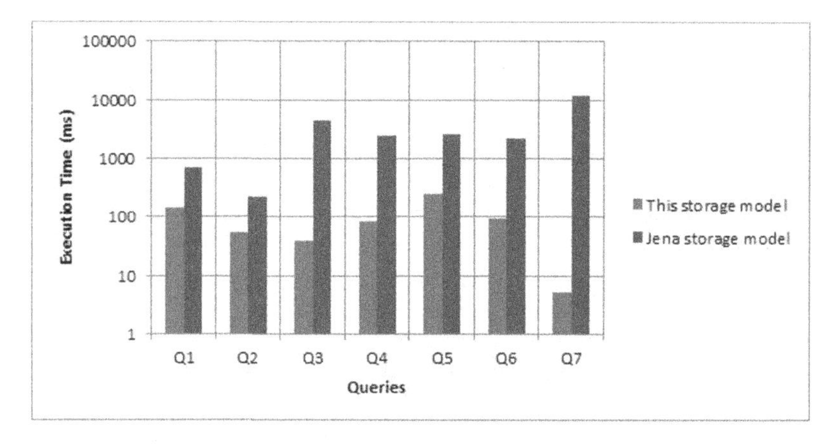

Fig. 10 An overview of the overall queries' performance for the standard dataset (1M)

Discussion: On the basis of experimentation, we can say that our proposed storage method shows an average of 84-89% improved performance (the size of the dataset, 12885 and 21725 reified statements) as compared to the Jena storage model. When running the query against the dataset of one million statements, the results of the experiment and the evaluation also improved by 93% or so in the performance compared with the Jena storage application. Queries 1 to 7 were particularly good in the performance. Thus, increasing the size of the dataset up to 1 million statements did not significantly alter its performance and still shows that our presented method substantially improves the query response times. Furthermore, by designing an efficient relational database model to store OWL data, through consideration of the

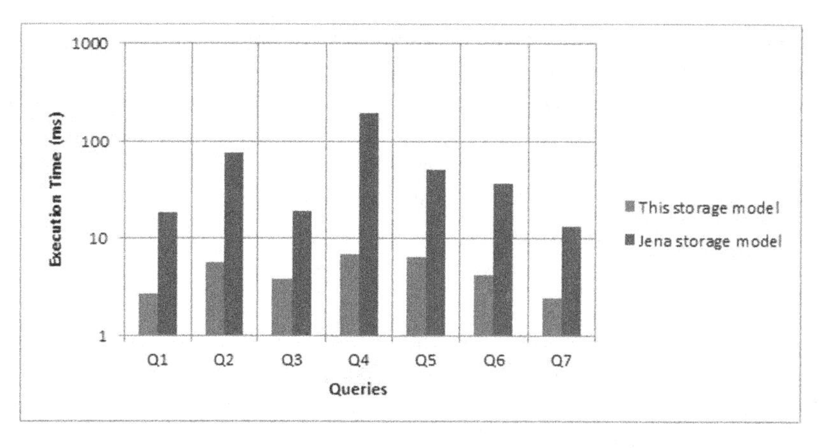

Fig. 11 An overview of the overall queries' performance for the university ontology dataset

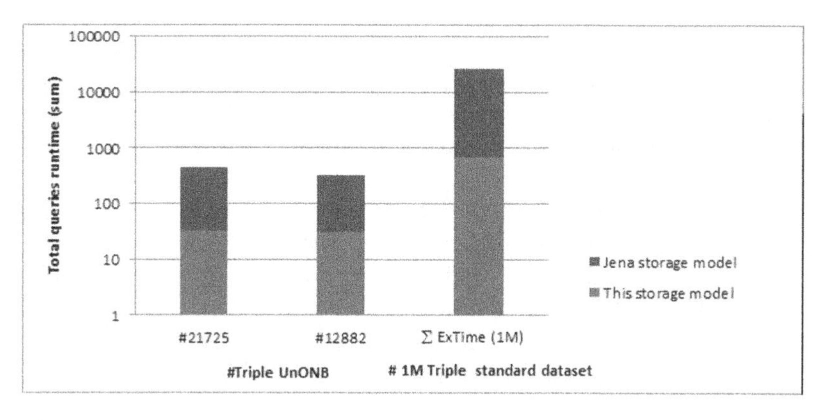

Fig. 12 Total outcomes for all queries

relationships between the structure and each component of the OWL documents, we found that our proposed model is efficient as it presents hierarchical knowledge among class or property where ontology classes, properties and instances are mapped to database schema with representing axioms, facts and restrictions. Most semantic triple stores based on relational database model use a single relational table; therefore it is not easy for users to understand the meanings of data. The proposed model, however, makes it very easy to comprehend data, as it is structured conveniently to enable understanding of meanings, as it provides hierarchical information in different tables. Our OWL storage model is also composed of well-defined rules to compensate for the drawbacks of earlier ontology storage techniques. While there are many solutions and tools for persisting ontologies in databases, most are a typical representative of the triple based storage systems. Another part of such methodologies is based on a sort of (partial) ontology meta-model. Furthermore, the proposed storage system schema is designed based on OWL 2 DL, which most regard as the full expressivity of OWL 2. Also, the proposed model stores the data completely and correctness as it answers, in entirety, different complex queries which require data from different sources. As well, it makes the data's management easy, due to managers have the capability of identifying semantically as well as systematically ontology in addition to storage structure.

6 Conclusion

The contributions of this paper are twofold. At first, we identified the problems with semantic storage system techniques for RDF and OWL. In the second part, we addressed the drawbacks of earlier techniques by proposing a new ontology storage technique superior to those previously used. We presented a novel approach to ontology persistence focused on the OWL 2 DL language. Our research aimed to improve the storage model of OWL by adopting several changes within the management of OWL data to design an efficient relational database model and store OWL ontology knowledge bases. We started from designing an efficient relational database layout that utilised an object relational mapping. We created a specific mapping with a set of process rules to store an ontology in a relational database. Results from our evaluation show that our storage method provides improved performance and stores the data completely and accurately as it answers the entire complex query, which requires data from different data sources. Moreover, we found that the proposed model is efficient as it presents hierarchical knowledge among class or property where ontology classes, properties and instances are mapped to database schema with representing axioms, facts and restrictions.

References

1. Astrova, I., Korda, N., Kalja, A.: Automatic transformation of OWL ontologies to SQL relational databases. In: IADIS European Conference Data Mining (MCCSIS), Lisbon, Portugal, vol. 3, pp. 145–149 (2007)
2. Baolin, L., Bo, H.: An evaluation of RDF storage systems for large data applications. In: First International Conference on Semantics, Knowledge and Grid, SKG 2005, pp. 59–59 (2005)
3. Bechhofer, S., van Harmelen, F., Hendler, J., Horrocks, I., McGuinness, D.L., Patel-Schneider, P.F., Stein, L.A.: OWL Web Ontology Language Reference. Technical report, W3C (February 2004), http://www.w3.org/TR/owl-ref/
4. Bechhofer, S. (ed.): Ontology language standardisation efforts. Technical report, IST Project, IST-2000-29243 (2002)
5. Berners-Lee, T., Hendler, J., Lassila, O.: Semantic web. Scientific American Magazine (May 2001)
6. Bizer, C., Schultz, A.: The Berlin SPARQL benchmark. International Journal on Semantic Web and Information Systems (2009)
7. Broekstra, J., Kampman, A., Harmelen, F.V.: Sesame: Storage and querying middleware for the semantic web (2008)

8. Calì, A., Gottlob, G., Lukasiewicz, T.: Datalog: a unified approach to ontologies and integrity constraints. In: Proceedings of the 12th International Conference on Database Theory, ICDT 2009, pp. 14–30. ACM, New York (2009)
9. Chen, Y., Ou, J., Jiang, Y., Meng, X.: HStar - A Semantic Repository for Large Scale OWL Documents. In: Mizoguchi, R., Shi, Z.-Z., Giunchiglia, F. (eds.) ASWC 2006. LNCS, vol. 4185, pp. 415–428. Springer, Heidelberg (2006)
10. Dalakleidi, K., Dasiopoulou, S., Stoilos, G., Tzouvaras, V., Stamou, G., Kompatsiaris, Y.: Semantic Representation of Multimedia Content. In: Paliouras, G., Spyropoulos, C.D., Tsatsaronis, G. (eds.) Multimedia Information Extraction. LNCS, vol. 6050, pp. 18–49. Springer, Heidelberg (2011)
11. Gali, A., Chen, C.X., Claypool, K.T., Uceda-Sosa, R.: From ontology to relational databases. In: ER 2004 Workshops, pp. 278–289 (2004)
12. Harrison, R., Chan, C.W.: Distributed ontology management system. In: Canadian Conference on Electrical and Computer Engineering, pp. 661–664 (2005)
13. Hitzler, P., Krötzsch, M., Parsia, B., Patel-Schneider, P.F., Rudolph, S. (eds.): OWL 2 Web Ontology Language: Primer. W3C Recommendation, October 27 (2009)
14. Horrocks, I.: Scalable ontology-based information systems. In: Proceedings of the 13th International Conference on Extending Database Technology, EDBT 2010, pp. 2–2. ACM, New York (2010)
15. Hu, W., Qu, Y.: Discovering simple mappings between relational database schemas and ontologies. In: Aberer, K., Choi, K.-S., Noy, N., Allemang, D., Lee, K.-I., Nixon, L.J.B., Golbeck, J., Mika, P., Maynard, D., Mizoguchi, R., Schreiber, G., Cudré-Mauroux, P. (eds.) ASWC 2007 and ISWC 2007. LNCS, vol. 4825, pp. 225–238. Springer, Heidelberg (2007)
16. Pulido, J.R.G., Ruiz, M.A.G., Herrera, R., Cabello, E., Legrand, S., Elliman, D.: Ontology languages for the semantic web: A never completely updated review. Knowledge-Based Systems 19(7), 489–497 (2006)
17. Smith, M., Welty, C., McGuinness, D.L.: OWL web ontology language guide. W3C recommendation (2004)
18. Stegmaier, F., Gröbner, U., Döller, M., Kosch, H., Baese, G.: Evaluation of current RDF database solutions. In: Proceedings of the 10th International Workshop on Semantic Multimedia Database Technologies, vol. 539, pp. 39–55 (2009)
19. Turhan, A.-Y.: Description logic reasoning for semantic web ontologies. In: Proceedings of the International Conference on Web Intelligence, Mining and Semantics, WIMS 2011, pp. 6:1–6:5. ACM, New York (2011)
20. W3C, Allegrograph rdfstore web 3.0's database (September 2009), `http://www.franz.com/agraph/allegrograph/`
21. Wood, D.: Kowari: A platform for semantic web storage and analysis. In: XTech 2005 Conference, pp. 05–0402 (2005)

DPF Workbench: A Diagrammatic Multi-Layer Domain Specific (Meta-)Modelling Environment

Yngve Lamo, Xiaoliang Wang, Florian Mantz, Wendy MacCaull, and Adrian Rutle

Abstract. This paper presents the DPF Workbench, a diagrammatic tool for domain specific modelling. The tool is an implementation of the basic ideas from the Diagram Predicate Framework (DPF), which provides a graph based formalisation of (meta)modelling and model transformations. The DPF Workbench consists of a specification editor and a signature editor and offers fully diagrammatic specification of domain-specific modelling languages. The specification editor supports development of metamodelling hierarchies with an arbitrary number of metalevels; that is, each model can be used as a metamodel for the level below. The workbench also facilitates the automatic generation of domain-specific specification editors out of these metamodels. Furthermore, the conformance relations between adjacent metalevels are dynamically checked by the use of typing morphisms and constraint validators. The signature editor is a new component that extends the DPF Workbench with functionality for dynamic definition of predicates. The syntax of the predicates are defined by a shape graph and a graphical icon, and their semantics are defined by validators. Those predicates are used to add constrains on the underlying graph. The features of the DPF Workbench are illustrated by a running example presenting a metamodelling hierarchy for workflow modelling in the health care domain.

1 Introduction

Model-driven engineering (MDE) promotes the use of models as the primary artefacts in the software development process. These models are used to specify, simulate, generate code and maintain the resulting applications. Models can be specified

Yngve Lamo · Xiaoliang Wang · Florian Mantz
Bergen University College, Norway
e-mail: {yla, xwa, fma}@hib.no

Wendy MacCaull · Adrian Rutle
St. Francis Xavier University, Canada
e-mail: {wmaccaul, arutle}@stfx.ca

R. Lee (Ed.): Computer and Information Science 2012, SCI 429, pp. 37–52.
springerlink.com

by general-purpose modelling languages such as the Unified Modeling Language (UML) [20], but to fully unfold the potentials of MDE, models are specified by Domain-Specific Modelling Languages (DSMLs), each tailored to a specific domain of concern [11]. DSMLs are modelling languages where the language primitives consist of domain concepts. Traditionally such domain concepts are specified by a graph based metamodel while the constraints are specified by a text based language such as the Object Constraint Language (OCL) [19]. This mixture of text based and graph based languages is an obstacle for employing MDE especially with regard to model transformations [25] and synchronisation of graphical models with their textual constraints [23]. A more practical solution to this problem is a fully graph based approach to the definition of DSMLs; i.e., diagrammatic specification of both the metamodel and the constraints [24].

The availability of tools that facilitate the design and implementation of DSMLs is an important factor for the acceptance and adoption of MDE. DSMLs are required to be intuitive enough for domain experts whereas they have a solid formal foundation which enables automatic verification and sound model transformations. Since DSMLs are defined by metamodels, these tools need to support automatic generation of specification editors out of metamodels.

An industrial standard language to describe DSMLs is the Meta-Object Facility (MOF) [18] provided by the Object Management Group (OMG). A reference implementation inspired by the MOF standard is Ecore, which is the core language of the Eclipse Modelling Framework (EMF) [26]. This framework uses a two-level metamodelling approach where a model created by the Ecore editor can be used to generate a DSML with a corresponding editor (see Fig. 1a). This editor, in turn, can be used to create instances; however, the instances of the DSML cannot be used to generate other DSMLs. That is, the metamodelling process is limited to only two user-defined metamodelling levels.

The two-level metamodelling approach has several limitations (see [13, 1] for a comprehensive argumentation). The lack of multi-layer metamodelling support

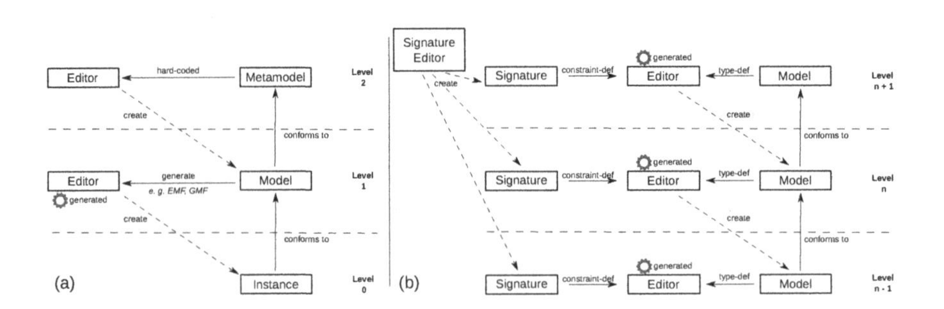

Fig. 1 A simplified view of (a) the EMF metamodelling hierarchy, and (b) a generic metamodelling hierarchy as implemented in the DPF Workbench

forces DSML designers to introduce type-instance relations in the metamodel. This leads to a mixture of domain concepts with language concepts in the same modelling level. The approach in this paper tackles this issue by introducing a multi-layer metamodelling tool.

This paper presents the DPF Workbench, a prototype diagrammatic (meta)modelling tool for the specification of diagrammatic signatures, (meta)models, and the generation of specification editors (see Fig. 1b). The DPF Workbench is an implementation of the techniques and methodologies developed in the Diagram Predicate Framework (DPF) [5], that provides a formalisation of (meta)modelling and model transformations based on category theory and graph transformations. The DPF Workbench supports the development of metamodelling hierarchies by providing an arbitrary number of metalevels; that is, each model at a metalevel can be used as a metamodel for the metalevel below. Moreover, the DPF Workbench checks the conformance of models to their metamodels by validating both typing morphisms and diagrammatic constraints. DPF Workbench extends the DPF Editor [15] with a signature editor which is used to define new domain specific predicates (syntax) and their corresponding validators (semantics).

The functionality of the DPF Workbench is demonstrated by specifying a metamodelling hierarchy for health workflows. Health services delivery processes are complicated and are frequently developed from regional or national guidelines which are written in natural language. Having good models of these processes is particularly valuable for several reasons (1) the modelling process which must be done in conjunction with (health) domain experts clarifies the meaning of the guidelines and has, in a number of situations, found ambiguities or inconsistencies in the guidelines; (2) graphical display of a process makes it easy for the (clinicians) domain experts to understand; (3) formal descriptions can be analysed for their behavioural characteristics via model checking; and, (4) the models can drive an executable workflow engine to guide the actual process in health care settings. The use of MDE technology is especially valuable because guidelines may be updated or changed every few years so the model and associated workflow must be redeveloped; moreover, though compliance with guidelines is required across a province or country, individual health districts, indeed individual hospitals or other service settings (clinics, etc.) will have processes that are specific to their setting so the overall process must be customised for the local setting. With MDE once the model is written and analysed for correctness the executable code is generated automatically. Moreover the abstraction required for development of abstract models makes it easier to involve domain experts in the development process. MDE transformation techniques can be used to generate code suitable for model checkers to verify behavioural characteristics of a workflow model, an important feature for a safety critical applications such as health care. Health care costs are rising dramatically worldwide; better outcomes for the patient as well as enhanced efficiencies have been shown to result from better process (workflow) definitions.

The remainder of the paper is organised as follows. Section 2 introduces some basic concepts from DPF. Section 3 gives a brief overview of the tool architecture. Section 4 demonstrates the functionality of the tool in a metamodelling scenario. Section 5 compares DPF Workbench with related tools, and finally Section 6 outlines future research and implementation work and concludes the paper.

2 Diagram Predicate Framework

In DPF, models are represented by *(diagrammatic) specifications*. A specification $\mathfrak{S} = (S, C^{\mathfrak{S}} : \Sigma)$ consists of an *underlying graph* S together with a set of *atomic constraints* $C^{\mathfrak{S}}$ [24, 23]. The graph represents the structure of the specification and the atomic constraints represent the restrictions attached to this structure. Atomic constraints are specified by *predicates* from a predefined *(diagrammatic) signature* Σ. A signature $\Sigma = (\Pi^{\Sigma}, \alpha^{\Sigma})$ consists of a collection of predicates, each having a symbol, an arity (or shape graph), a visualisation and a semantic interpretation (see Table 1).

Table 1 The signature Σ used in the metamodelling example

Π^{Σ}	$\alpha^{\Sigma}(\pi)$	**Visualisation**	**Semantics**
[mult(m,n)]	$1 \xrightarrow{a} 2$	$X \xrightarrow{f} Y$ [m..n]	$\forall x \in X : m \leq \lvert f(x) \rvert \leq n$, with $0 \leq m \leq n$ and $n \geq 1$
[irreflexive]	1 with loop a	X with loop f [irr]	$\forall x \in X : x \notin f(x)$
[injective]	$1 \xrightarrow{a} 2$	$X \xrightarrow{f} Y$ [inj]	$\forall x, x' \in X : f(x) = f(x')$ implies $x = x'$
[nand]	$1 \xrightarrow{a} 2$, $1 \xrightarrow{b} 3$	$X \xrightarrow{f} Y$, $X \xrightarrow{g} Z$ [nand]	$\forall x \in X :$ $f(x) = \emptyset \vee g(x) = \emptyset$
[surjective]	$1 \xrightarrow{a} 2$	$X \xrightarrow{f} Y$ [surj]	$f(X) = Y$
[jointly-surjective_2]	$1 \xrightarrow{a} 2$, $3 \xrightarrow{g} 2$	$X \xrightarrow{f} Y$, $Z \xrightarrow{b} Y$ [js]	$f(X) \cup g(Z) = Y$
[xor]	$1 \xrightarrow{a} 2$, $1 \xrightarrow{b} 3$	$X \xrightarrow{f} Y$, $X \xrightarrow{g} Z$ [xor]	$\forall x \in X :$ $(f(x) = \emptyset \vee g(x) = \emptyset)$ and $(f(x) \neq \emptyset \vee g(x) \neq \emptyset)$

In the DPF Workbench, a DSML corresponds to a specification editor, which in turn consists of a signature and a metamodel. A specification editor can be used to specify new metamodels, and thus define new DSMLs (see Fig. 1b).

Next we show a specification Fig. 2 is an example of a specification $\mathfrak{S}_2$ that ensures that "activities cannot send messages to themselves". In $\mathfrak{S}_2$, this requirement is forced by the atomic constraint $(\texttt{[irreflexive]}, \delta)$ on the arrow Message. Note that δ is a graph homomorphism $\delta : (\, 1 \circlearrowleft^{a} \,) \to (\, \mathsf{Activity} \circlearrowleft^{\mathsf{Message}} \,)$ specifying the part of $\mathfrak{S}_2$ to which the [irreflexive] predicate is added.

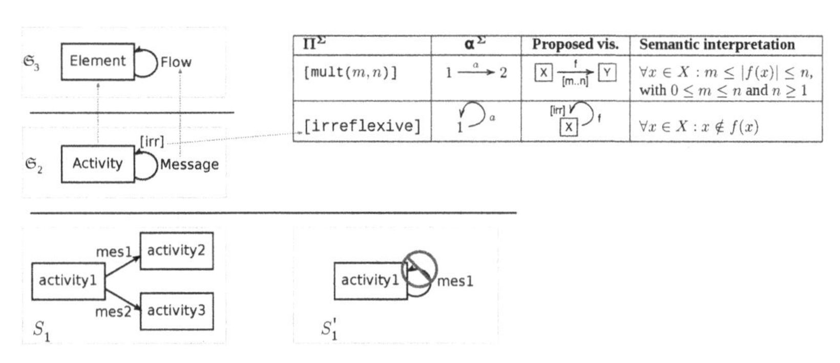

Π^{Σ}	α^{Σ}	Proposed vis.	Semantic interpretation
[mult(m,n)]	$1 \xrightarrow{a} 2$	$\boxed{X} \xrightarrow[\text{[m..n]}]{f} \boxed{Y}$	$\forall x \in X : m \leq \lvert f(x) \rvert \leq n$, with $0 \leq m \leq n$ and $n \geq 1$
[irreflexive]	$1 \circlearrowleft^{a}$	[irr] $\boxed{X} \circlearrowleft^{f}$	$\forall x \in X : x \notin f(x)$

Fig. 2 The specifications $\mathfrak{S}_2$, $\mathfrak{S}_3$, the signature Σ, a valid instance S_1 of $\mathfrak{S}_2$, and an invalid instance S_1' of $\mathfrak{S}_2$ that violates the irreflexivity constraint

The semantics of the underlying graph of a specification must be chosen in a way that is appropriate for the corresponding modelling environment [24, 23]. In object-oriented structural modelling, each object may be related to a set of other objects. Hence, it is appropriate to interpret nodes as sets and arrows $X \xrightarrow{f} Y$ as multi-valued functions $f : X \to \wp(Y)$. The powerset $\wp(Y)$ of Y is the set of all subsets of Y; i.e. $\wp(Y) = \{A \mid A \subseteq Y\}$. Moreover, the composition of two multi-valued functions $f : X \to \wp(Y)$, $g : Y \to \wp(Z)$ is defined by $(f;g)(x) := \bigcup\{g(y) \mid y \in f(x)\}$.

The semantics of a specification is defined by the set of its instances (I, ι) [9]. An instance (I, ι) of $\mathfrak{S}$ is a graph I together with a graph homomorphism $\iota : I \to S$ that satisfies the atomic constraints $C^{\mathfrak{S}}$. To check that an atomic constraint is satisfied in a given instance of $\mathfrak{S}$, it is enough to inspect the part of S that is affected by the atomic constraint [23]. In this way, an instance of the specification is inspected first to check that the typing is correct, then to check that every constraint in the specification is satisfied. For example, Fig. 2 shows two graphs S_0, S_0', both typed by the specification $\mathfrak{S}_2$, but only S_0 is a valid instance of $\mathfrak{S}_2$, since S_0' violates the $(\texttt{[irreflexive]}, \delta)$ constraint on Message by having a circular (reflexive) arrow of type Message.

In DPF, two kinds of conformance relations are distinguished: *typed by* and *conforms to*. A specification $\mathfrak{S}_i$ at metalevel i is said to be typed by a specification $\mathfrak{S}_{i+1}$ at metalevel $i+1$ if there exists a graph homomorphism $\iota_i : S_i \to S_{i+1}$, called the typing morphism, between the underlying graphs of the specifications. A specification $\mathfrak{S}_i$ at metalevel i is said to conform to a specification $\mathfrak{S}_{i+1}$ at metalevel $i+1$ if there exists a typing morphism $\iota_i : S_i \to S_{i+1}$ such that (S_i, ι_i) is a valid instance of $\mathfrak{S}_{i+1}$; i.e. such that ι_i satisfies the atomic constraints $C^{\mathfrak{S}_{i+1}}$.

For instance, Fig. 2 shows a specification $\mathfrak{S}_2$ that conforms to a specification $\mathfrak{S}_3$. That is, there exists a typing morphism $\iota_2 : S_2 \to S_3$ such that (S_2, ι_2) is a valid instance of $\mathfrak{S}_3$. Note that since $\mathfrak{S}_3$ does not contain any atomic constraints, the underlying graph of $\mathfrak{S}_2$ is a valid instance of $\mathfrak{S}_3$ as long as there exists a typing morphism $\iota_2 : S_2 \to S_3$.

3 Tool Architecture

The DPF Workbench has been developed in Java as a plug-in for Eclipse [10]. Eclipse follows a cross-platform architecture that is well suited for tool integration since it implements the Open Services Gateway initiative framework (OSGi). Moreover, it has an ecosystem around the basic tool platform that offers a rich set of plug-ins and APIs that are helpful when implementing modelling tools. In addition, Eclipse technologies are widely used in practise and are also employed in commercial products such as the Rational Software Architect (RSA) [14] as well as in open-source products such as the modelling tool TOPCASED [27]. For this reason the DPF Workbench can be integrated into such tools easily and used together with them.

Figure 3 illustrates that the DPF Workbench basically consists of three components (Eclipse plugins). The bottom component (the Core Model Management Component) provides the core features of the tool: these are the facilities to create, store and validate DPF specifications. This part uses EMF for data storage. This means the DPF Workbench contains an internal metamodel that is an Ecore model. As a consequence, each DPF specification is also an instance of this internal metamodel. EMF has been chosen for data storage since it is a de facto standard in the modelling field and guarantees high interoperability with various other tools and frameworks. Therefore, DPF models can be used with e.g., code generation frameworks such as those offered by the Eclipse Model To Text (M2T) project. Recently an adapter for Xpand (M2T) has been added to the workbench offering native support for DPF specifications. This means Xpand templates can use the DPF metamodelling hierarchy in addition to the one which is given by EMF.

The top component (the Visual Component) provides the visual editors, i.e., specification editors and signature editors. This component is implemented using the Graphical Editing Framework (GEF). GEF provides technology to create rich graphical editors and views for the Eclipse Workbench. The component mainly consists of classes following GEF's Model-View-Controller (MVC) architecture. There are Figures classes (constituting the view), Display Model classes and controller classes (named Parts in accordance with the GEF terminology). Special arrow-routing and display functions have been developed for showing DPF's special kinds of predicates.

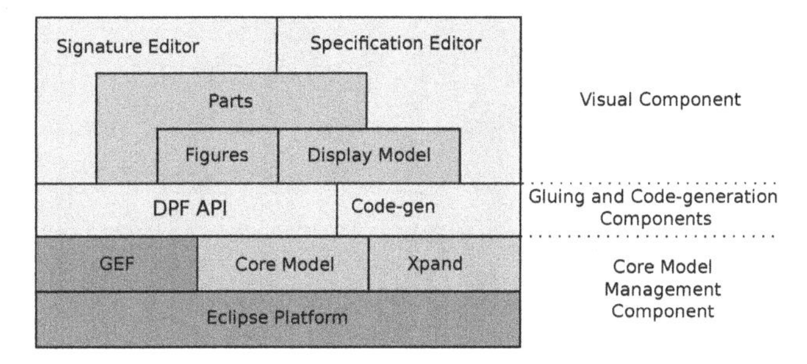

Fig. 3 The main component architecture of the DPF Workbench plug-in packages

The middle component (the Gluing Component) is used as mediator between the first two components. It ties together the functionality and manages file loading, object instantiation and general communication with the Eclipse platform.

4 A Metamodelling Example

This section illustrates the steps of specifying a metamodelling hierarchy using the DPF Workbench. The example demonstrates a metamodelling hierarchy that is used to specify a workflow for the treatment of Cancer Related Pain [6]. First we use the signature editor to define the signature that is used in the example. Then we show how to specify a metamodel using the DPF Workbench. We also show the generation of specification editors for DSML by loading an existing metamodel and an existing signature into the tool. Furthermore we present how type checking and constraint validation are performed by the workbench.

4.1 Creating Signatures

We first explain how we create a new project in the DPF Workbench and define the signature; i.e., the predicates that will be available in the modelling process. The DPF Workbench runs inside Eclipse, and has its own custom perspective. To get started, we activate the specification editor by selecting a project folder and invoking an Eclipse-type wizard for creating a new DPF Signature. The signature for the metamodelling hierarchy must include the predicates from Table 1. We use the signature editor to define the arity of the predicates, a graphical icon that illustrates the predicates in the DPF Workbench toolbar and the semantics of the predicates. Figure 4 shows how the arity of the `[xor]` predicate is defined. Figure 5 shows how the semantics of the `[xor]` predicate is defined as a Java validator. An example usage of the `[xor]` predicate explained in Section 4.2. Currently the semantics can only be defined by Java validators, but in future, it will be possible to define the semantics also by use of OCL syntax.

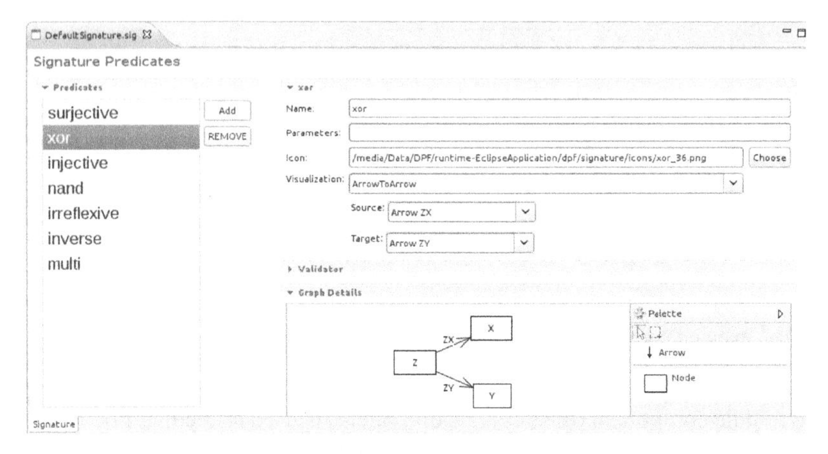

Fig. 4 Definition of the arity and the graphical icon of the `[xor]` predicate

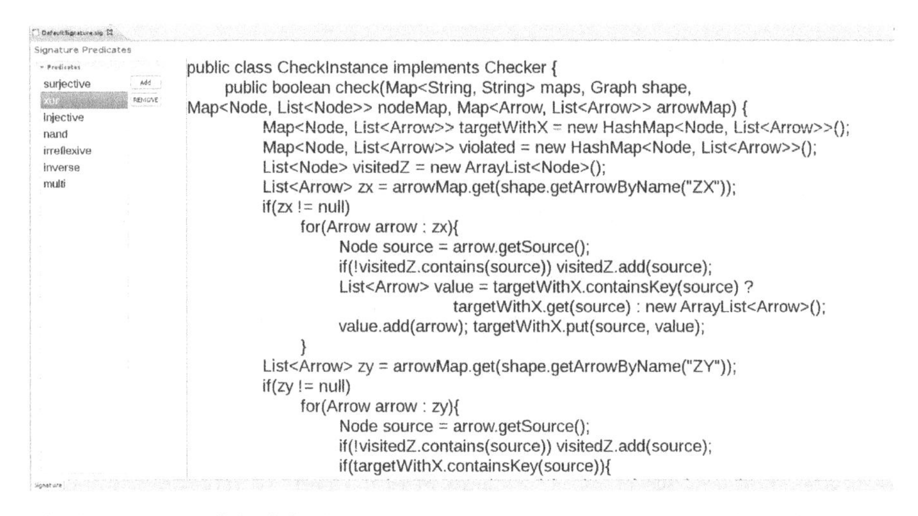

Fig. 5 An excerpt of the definition of the Java validator for the `[xor]` predicate

4.2 Defining Metamodels

After defining all the necessary predicates we load the DPF Workbench with the desired set of predicates corresponding to the signature shown in Table 1.

We start the metamodelling process by configuring the tool with the DPF Workbench's default metamodel $\mathfrak{S}_4$ consisting of Node and Arrow, that serves as a starting point for metamodelling in the DPF Workbench. This default metamodel is used as the type graph for the metamodel $\mathfrak{S}_3$ at the highest level of abstraction of the workflow metamodelling hierarchy. In $\mathfrak{S}_3$, we introduce the domain concepts Elements and Control, that are typed by Node (see Fig. 6). We also introduce Flow,

NextControl, ControlIn and ControlOut, that are typed by Arrow. The typing of this metamodel by the default metamodel is guaranteed by the fact that the tool allows only creation of specifications in which each specification element is typed by Node or by Arrow. One requirement for process modelling is that "each control should have at least one incoming arrow from an element or another control"; this is specified by adding the [jointly-surjective_2] constraint on the arrows ControlIn and NextControl. Another requirement is that "each control should be followed by either another control or by an element, not both"; this is specified by the [xor] constraint on the arrows ControlOut and NextControl. We save this specification in a file called process_m3.dpf, with "m3" reflecting the metalevel M_3 to which it belongs.

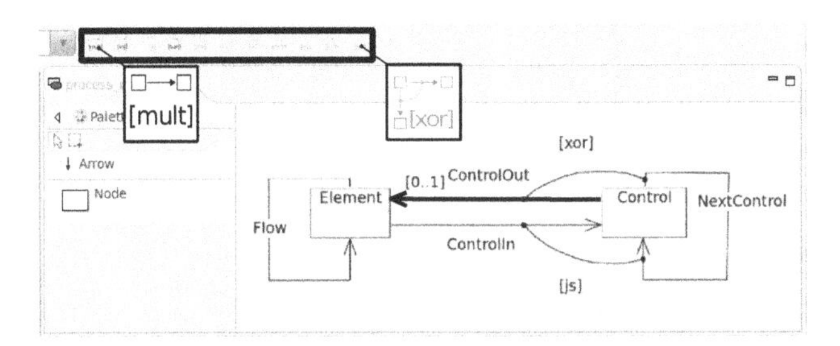

Fig. 6 DPF Workbench configured with the default metamodel consisting of Node and Arrow, and the signature Σ from Table 1 indicated with a bold black rectangle; showing also the specification $\mathfrak{S}_3$ under construction; note that the bold black arrow ControlOut is selected, therefore the predicates that have $1 \rightarrow 2$ as their arity are enabled in the signature bar.

4.3 Generating Specification Editors from Metamodels

In this section, we illustrate how a specification editor can be generated from the existing specification $\mathfrak{S}_3$. This is achieved by again invoking the wizard for creating a new DPF Specification Diagram. This time, in addition to specifying that our file shall be called process_m2.dpf, we also specify that the file process_m3.dpf shall be used as the metamodel for our new specification $\mathfrak{S}_2$. We use the signature from Table 1 with this new specification editor. Note that the tool palette in Fig. 7 contains buttons for each specification element defined in Fig. 6. In process_m2.dpf we will define a specification $\mathfrak{S}_2$ which is compliant with the following requirements:

1. Each *activity* may send *messages* to one or more *activities*
2. Each *activity* may be *sequenced* to another *activity*
3. Each *activity* may be connected to at most one *choice*
4. Each *choice* must be connected to at least two *conditions*
5. Each *activity* may be connected either to a *choice* or to another *activity*, but not both.
6. Exactly one *activity* must be connected to each *choice*

7. Each *condition* must be connected to exactly one *activity*
8. An *activity* cannot send messages to itself
9. An *activity* cannot be sequenced to itself

We will explain now how some of the requirements above are specified in $\mathfrak{S}_2$. The requirements 1 and 2 are specified by introducing the Activity node that is typed by Element, as well as Message and Sequence arrows that are typed by Flow. The requirement 5 is specified by adding the constraint `[nand]` on the arrows Sequence and Choice. The requirement 6 is specified by adding the constraints `[injective]` and `[surjective]` on ChoiceIn. The requirements 8 and 9 are specified by adding the constraint `[irreflexive]` on Message and Sequence, respectively.

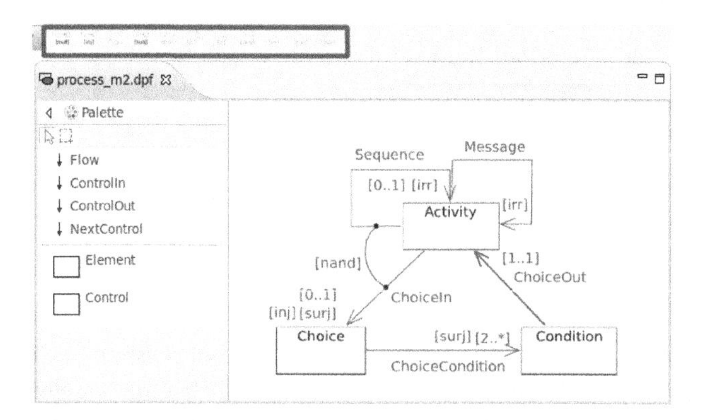

Fig. 7 The DPF Workbench configured with the specification $\mathfrak{S}_3$ from Fig. 6 as metamodel, and the signature Σ from Table 1 indicated with a bold black rectangle; the specification $\mathfrak{S}_2$ under construction is also shown.

4.4 Conformance Checks

The conformance relation between $\mathfrak{S}_2$ and $\mathfrak{S}_3$ is checked in two steps. Firstly, the specification $\mathfrak{S}_2$ is correctly typed over its metamodel by construction. The DPF Workbench actually checks that there exists a graph homomorphism from the specification to its metamodel while creating the specification. For instance, when we create the ChoiceIn arrow of type ControlIn, the tool ensures that the source and target of ChoiceIn are typed by Element and Control, respectively. Secondly, the constraints are checked by corresponding validators during creation of specifications. In Fig. 7 we see that all constraints specified in $\mathfrak{S}_3$ are satisfied by $\mathfrak{S}_2$. However, Fig. 8 shows a specification which violates some of the constraints of $\mathfrak{S}_3$, e.g., the `[xor]` constraint on the arrows ControlOut and NextControl in $\mathfrak{S}_3$ is violated by the arrow WrongArrow in $\mathfrak{S}_2$. The constraint is violated since Condition – that is typed by Control – is followed by both a Choice, and an Activity, violating the requirement "each control should be followed by either another control or by an element, not both". This violation will be indicated in the tool by a message (or a tip) in the status bar.

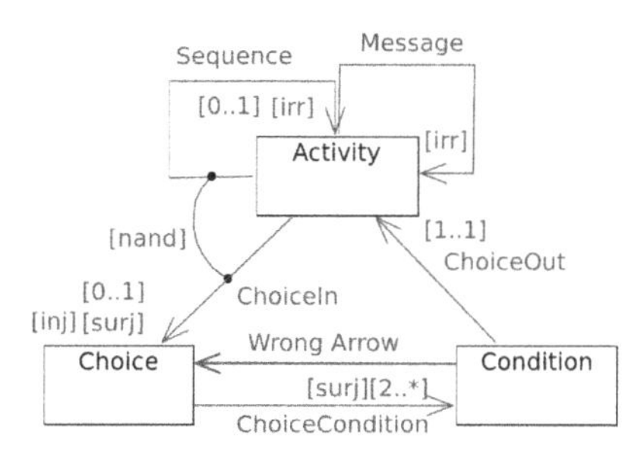

Fig. 8 A specification violating the `[xor]` constraint on the arrows ControlOut and NextControl in $\mathfrak{S}_3$

4.5 Further Modelling Layers

We can now repeat the previous step and load the specification editor with the specification $\mathfrak{S}_2$ (by choosing `process_m2.dpf`) as metamodel. This editor is then used to specify the workflow model at the metalevel M_1. The example is taken from the Guidelines for the Management of Cancer-Related Pain in Adults [6]. This guideline outlines the procedure to manage a patient's pain. The Pain Assessment activity assesses all causes of pain (total pain), determine pain location(s), pain intensity, and other symptoms. Complete history and all previous analgesics (including opioids) and response to each will be documented in this activity. After assessment, if the patient is currently under any opioid medication then the flow will be either forwarded to the Strong Opioid Regimen1 or Strong Opioid Regimen2 activity, depending on the pain level and current opioid dose. Otherwise, the flow goes to the Non-Opioid or Weak Opioid or Strong Opioid Regimen1 activity. While a patient is taking any Strong Opioid medication, his/her pain intensity is assessed regularly and the dose is adjusted accordingly. If any symptoms for opioid toxicity or other side effects are found, they are managed appropriately. We modelled the workflow in the DPF Editor and ensured that the model is conformant to its metamodel. In [21], the authors modelled the guideline using a different workflow modelling language (called CWML). They monitored some interesting properties involving pain reassessment times and verified some behavioural LTL-properties using an automated translator to a model checker. In that work, the authors did not focus on metamodelling or the conformance aspects.

The guideline is represented as the DPF specification $\mathfrak{S}_1$ in Fig. 9. Note that this time the tool palette contains buttons for each specification element defined in Fig. 7. For this tool palette (not shown in Fig. 9) we have chosen a concrete syntax for process modelling with special visual effects for model elements. For instance, model elements typed by Choice and Condition are visualised as diamonds and circles, respectively. In future, to enhance readability, the specification editor will facilitate other visualisation functionalities like zooming, grouping, etc.

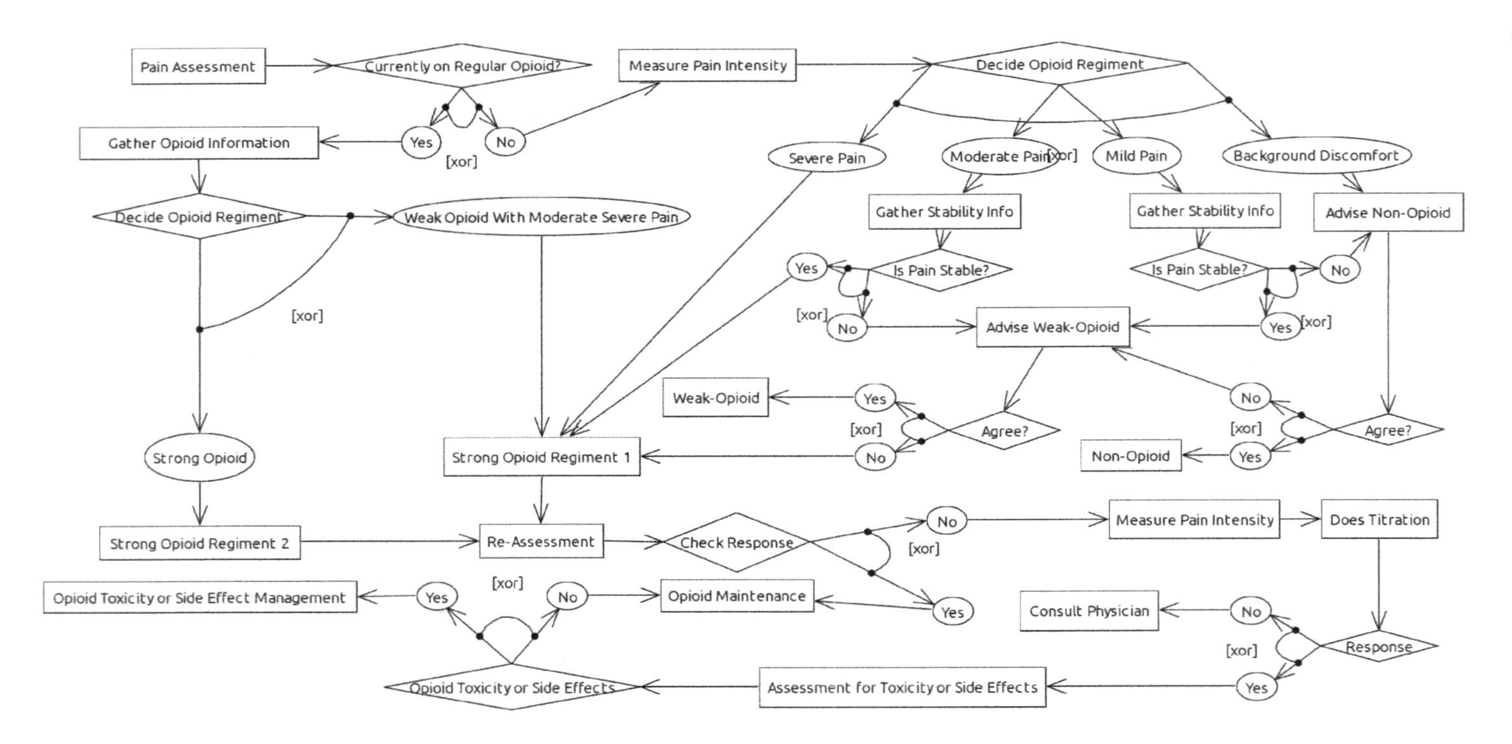

Fig. 9 The Guidelines for the Management of Cancer-Related Pain in Adults

Finally, we may use predicates from the signature to add constraints to $\mathfrak{S}_1$, and, we may use $\mathfrak{S}_1$ as a metamodel for another modelling level. This pattern could be repeated as deep as it is necessary, however, in this example we stop at this level, and will eventually generate the code that is used for model checking of $\mathfrak{S}_1$.

5 Related Work

There is an abundance of visual modelling tools available, both as open-source software and as closed-source commercial products. Some of these tools also possess metamodelling features, letting the users specify a metamodel and then use this metamodel to create a new specification editor. Table 2 summaries the comparison of some popular metamodelling tools with the DPF Workbench.

Table 2 Comparison of the DPF Workbench to other metamodelling tools, EVL stands for Epsilon Validation Language, and the current predefined validator in DPF is implemented in Java

Tool	No. of Layers	Diagrammatic language	Constraint	Platform	Visual UI
EMF/GMF	2		OCL, EVL, Java	Java VM	✓
VMTS	∞		OCL	Windows	✓
AToM^3	2		OCL, Python	Python, Tk/tcl	✓
GME	2		OCL	Windows	✓
metaDepth	∞		EVL	Java VM	
DPF	∞	✓	Predefined validator	Java VM	✓

The Visual Modelling and Transformation System (VMTS) supports editing models according to their metamodels [17]. AToM^3 (A Tool for Multi-formalism and Meta-Modelling) is a tool for multi-paradigm modelling [2, 8]; formalisms and models are described as graphs. From the metamodel of a formalism, AToM^3 can generate a tool that lets the user create and edit models described in the specified formalism. Some of the metamodels currently available are: Entity-Relationship, Deterministic and Non-Deterministic Finite State Automata, Data Flow Diagrams, etc.

The Generic Modelling Environment (GME) [16] is a configurable toolkit for creating domain-specific modelling and program synthesis environments. The configuration is accomplished through metamodels specifying the modelling paradigm (modelling language) of the application domain [12]. The GME metamodelling language is based on the UML class diagram notation and OCL constraints.

The metaDepth [7] framework permits building systems with an arbitrary number of metalevels through deep metamodelling. The framework allows the specification and evaluation of derived attributes and constraints across multiple metalevels, linguistic extensions of ontological instance models, transactions, and hosting different constraint and action languages. At present, the framework supports only textual specifications; however, there is some work in progress on integrating DPF with metaDepth that aims to give a graph based formalisation of metaDepth, and deep metamodelling in general.

The table shows that VMTS, metaDepth and DPF Workbench support n-layer metamodelling, while other three tools, AToM^3, GME and EMF/GMF only support two level metamodelling. Most tools use OCL as their constraint language while Java, EVL and Python are alternatives. Those tools have no support for diagrammatic constraints, except the DPF Workbench, which has a dynamic definition of constraint syntax and corresponding semantics by use of Java validators.

6 Conclusion and Future Work

In this paper, we presented the prototype (meta)modelling tool DPF Workbench. The tool is developed in Java and runs as a plug-in on the Eclipse platform. The DPF Workbench supports fully diagrammatic metamodelling as proposed by the DPF Framework. The functionality of the tool has been illustrated by specifying a metamodelling hierarchy for health workflow modelling. Workflow is becoming an increasingly popular paradigm across many domains to articulate process definitions and guide actual processes. While the methods discussed here have been applied to a case study involving workflow for a health care system, they can be used to develop correct and easily customisable executable process definitions for many complex and safety critical systems. It has been shown how the specification editor's tool palette can be configured for a given domain by using a specific metamodel and a specific signature. To ensure correct typing of the edited models the tool uses graph homomorphisms. Moreover, it implements a validation mechanism that checks instances against all the constraints that are specified by the metamodel. We have also shown how models created in the tool can be used as metamodels at an arbitrary number of metamodelling levels. The authors are not aware of other EMF based tools that facilitate multi-level metamodelling. The DPF Workbench also includes a signature editor to dynamically create new predicates and their corresponding semantics.

Many directions for further work still remains unexplored, other are currently in the initial development phases. We shall only mention the most prominent here:

Code generation. The real utility for an end user of DPF Workbench will become manifest when an actual running system can be generated from specifications. We have already done some introductory work on code generation [4], and lately an Xpand adapter is included to the DPF Workbench. The adapter will be used in a further work to automatically generate scripts that will be used for model checking workflows specified by the DPF Workbench.

Configurable concrete syntax. As the system exists today, all diagram (nodes, arrows and constraints) visualisations are hardcoded in the specification editor code. A desirable extension would be to make visualisations more decoupled from the rest of the Display Model than is the current situation. This would involve a configurable and perhaps directly editable *concrete syntax* [3].

Layout and routing. Automated layout seems to become an issue when dealing with medium-sized to large diagrams. There seems to be a big usability gain to be capitalised on in this matter. Today's specification editor contains a simple routing algorithm, based on GEF's `ShortestPathConnectionRouter` class. The problem of finding routing algorithms that produce easy-readable output is a focus of continuous research [22], and this problem applied to DPF Workbench can probably be turned into a separate research task.

In addition to these areas, development to utilise the core functionality of DPF Workbench as a base for model transformation and (meta)model evolution is on the horizon, reflecting the theoretical foundations that are being laid down within the DPF research community.

References

1. Atkinson, C., Kühne, T.: Rearchitecting the UML infrastructure. ACM Transactions on Modeling and Computer Simulation 12(4), 290–321 (2002), doi:10.1145/643120.643123
2. AToM^3: A Tool for Multi-formalism and Meta-Modelling: Project Web Site, `http://atom3.cs.mcgill.ca/`
3. Baar, T.: Correctly Defined Concrete Syntax for Visual Modeling Languages. In: Wang, J., Whittle, J., Harel, D., Reggio, G. (eds.) MoDELS 2006. LNCS, vol. 4199, pp. 111–125. Springer, Heidelberg (2006)
4. Bech, Ø., Lokøen, D.V.: DPF to SHIP Validator Proof-of-Concept Transformation Engine, `http://dpf.hib.no/code/transformation/dpf_to_shipvalidator.py`
5. Bergen University College and University of Bergen: Diagram Predicate Framework Web Site, `http://dpf.hib.no/`
6. Broadfield, L., Banerjee, S., Jewers, H., Pollett, A., Simpson, J.: Guidelines for the Management of Cancer-Related Pain in Adults. Supportive Care Cancer Site Team, Cancer Care Nova Scotia (2005)
7. de Lara, J., Guerra, E.: Deep Meta-modelling with METADEPTH. In: Vitek, J. (ed.) TOOLS 2010. LNCS, vol. 6141, pp. 1–20. Springer, Heidelberg (2010)
8. de Lara, J., Vangheluwe, H.: Using AToM^3 as a Meta-CASE Tool. In: Proceedings of ICEIS 2002: 4th International Conference on Enterprise Information Systems, Ciudad Real, Spain, pp. 642–649 (2002)
9. Diskin, Z., Wolter, U.: A Diagrammatic Logic for Object-Oriented Visual Modeling. In: Proceedings of ACCAT 2007: 2nd Workshop on Applied and Computational Category Theory, vol. 203(6), pp. 19–41. Elsevier (2008), doi:10.1016/j.entcs.2008.10.041
10. Eclipse Platform: Project Web Site, `http://www.eclipse.org`
11. Fowler, M.: Domain-Specific Languages. Addison-Wesley Professional (2010)
12. GME: Generic Modeling Environment: Project Web Site, `http://www.isis.vanderbilt.edu/Projects/gme/`
13. Gonzalez-Perez, C., Henderson-Sellers, B.: Metamodelling for Software Engineering. Wiley (2008)

14. IBM: Rational Software Architect, http://www-01.ibm.com/software/awdtools/architect/swarchitect/
15. Lamo, Y., Wang, X., Mantz, F., Bech, Ø., Rutle, A.: DPF Editor: A Multi-Layer Diagrammatic (Meta)Modelling Environment. In: Proceedings of SPLST 2011: 12th Symposium on Programming Languages and Software (2011)
16. Ledeczi, A., Maroti, M., Bakay, A., Karsai, G., Garrett, J., Thomason, C., Nordstrom, G., Sprinkle, J., Volgyesi, P.: The Generic Modeling Environment. In: Proceedings of WISP 2001: Workshop on Intelligent Signal Processing, vol. 17, pp. 82–83. ACM (2001), http://www.isis.vanderbilt.edu/sites/default/files/GME2000Overview.pdf
17. Lengyel, L., Levendovszky, T., Charaf, H.: Constraint Validation Support in Visual Model Transformation Systems. Acta Cybernetica 17(2), 339–357 (2005)
18. Object Management Group: Meta-Object Facility Specification (2006), http://www.omg.org/spec/MOF/2.0/
19. Object Management Group: Object Constraint Language Specification (2010), http://www.omg.org/spec/OCL/2.2/
20. Object Management Group: Unified Modeling Language Specification (2010), http://www.omg.org/spec/UML/2.3/
21. Rabbi, F., Mashiyat, A.S., MacCaull, W.: Model checking workflow monitors and its application to a pain management process. In: Proceedings of FHIES 2011: 1st International Symposium on Foundations of Health Information Engineering and Systems, pp. 110–127 (2011), http://www.iist.unu.edu/ICTAC/FHIES2011/Files/fhies2011_8_17.pdf
22. Reinhard, T., Seybold, C., Meier, S., Glinz, M., Merlo-Schett, N.: Human-Friendly Line Routing for Hierarchical Diagrams. In: Proceedings of ASE 2006: 21st IEEE/ACM International Conference on Automated Software Engineering, pp. 273–276. IEEE Computer Society (2006)
23. Rutle, A.: Diagram Predicate Framework: A Formal Approach to MDE. Ph.D. thesis, Department of Informatics, University of Bergen, Norway (2010)
24. Rutle, A., Rossini, A., Lamo, Y., Wolter, U.: A Diagrammatic Formalisation of MOF-Based Modelling Languages. In: Oriol, M., Meyer, B. (eds.) TOOLS EUROPE 2009. LNBIP, vol. 33, pp. 37–56. Springer, Heidelberg (2009)
25. Rutle, A., Rossini, A., Lamo, Y., Wolter, U.: A Formalisation of Constraint-Aware Model Transformations. In: Rosenblum, D.S., Taentzer, G. (eds.) FASE 2010. LNCS, vol. 6013, pp. 13–28. Springer, Heidelberg (2010)
26. Steinberg, D., Budinsky, F., Paternostro, M., Merks, E.: EMF: Eclipse Modeling Framework 2.0, 2nd edn. Addison-Wesley Professional (2008)
27. TOPCASED: Project Web Site, http://www.topcased.org

Belief Revision for Intelligent Web Service Recommendation

Raymond Y.K. Lau and Long Song

Abstract. With the increasing number of web services deployed to the world wide web these days, discovering, recommending, and invoking web services to fulfil the specific functional and preferential requirements of a service user has become a very complex and time consuming activity. Accordingly, there is a pressing need to develop intelligent web service discovery and recommendation mechanisms to improve the efficiency and effectiveness of service-oriented systems. The growing interests in semantic web services has highlighted the advantages of applying formal knowledge representation and reasoning models to raise the level of autonomy and intelligence in human-to-machine and machine-to-machine interactions. Although classical logics such as description logic underpinning the development of OWL has been explored for services discovery, services choreography, services enactment, and services contracting, the non-monotonicity in web service discovery and recommendation is rarely examined. The main contribution of this paper is the development of a belief revision logic based service recommendation agent to address the non-monotonicity issue of service recommendation. Our initial experiment based on real-world web service recommendation scenarios reveals that the proposed logical model for service recommendation agent is effective. To the best of our knowledge, the research presented in this paper represents the first successful attempt of applying belief revision logic to build adaptive service recommendation agents.

Keywords: Web Services, Semantic Web, Services Recommendation, Belief Revision, Intelligent Agents.

1 Introduction

The emerge of web services facilitates the interoperability among computing platforms over the Internet [21, 22, 25]. In addition, the vision of Semantic Web raises

Raymond Y.K. Lau · Long Song
Department of Information Systems, City University of Hong Kong, Hong Kong SAR
e-mail: raylau@cityu.edu.hk, longsong@student.cityu.edu.hk

R. Lee (Ed.): Computer and Information Science 2012, SCI 429, pp. 53–66.
springerlink.com

the level of automation over web-based services invocation, and enhances not just machine-to-machine interactions, but also human-machine interactions [2]. The recent research into the Semantic Web has led to the development of emerging standards such as OWL, WSMO, and SWSL. More specifically, the wide spread interests of applying formal models (e.g., logic-based knowledge representation and reasoning) to develop various core components related to semantic web services has been observed [4, 8, 18, 19, 24]. In the context of human-machine service interactions, the increasing deployment and invocation of web services on the Internet has posed the new research challenges such as intelligent discovery and recommendation of optimal web services that best meet the users' preferences and their expectation about service quality [11, 29].

Quality-of-Service (QoS) in web services considers a service's non-functional characteristics (e.g., response time, failure rate, etc.) during service specification, discovery and composition. In order to facilitate the development of QoS-aware web services, a QoS-aware model usually takes into account a set of QoS attributes such as response time, throughput, reliability, availability, price, and so on [11, 29]. In this paper, we argue that both QoS and personalized service preferences should be taken into account in the context of intelligent web service recommendation. In addition, following the semantic web service paradigm for maximal machine-to-machine and human-to-machine interactions [2, 8, 24], a formal knowledge representation and reasoning model is preferred. Given the rapidly increasing number of web services deployed to the Internet these days, discovering, recommending, and invoking web services to fulfil the functional requirements and personal preferences of a user, has become a very complex and time consuming activity. One possible solution to alleviate such a problem which hinders the wide spread adoption of semantic web service is to develop a sound and effective recommendation (matching) mechanism to autonomously deduce the best services fulfilling the specific functional and personal requirements of the user.

For research in semantic web services, classical logic-based models have been applied to services discovery [24], services choreography [19], services enactment [8], and services contracting [18]. However, classical logics such as the description logic which underpins the development of OWL [3] adopt a very strict binary inference relation ($\vdash$). For instance, a web service can only be fully satisfying the service selection requirements of a user or not. Unfortunately, this is often not the case for most real-world web service invocation scenarios. A web service may only partially satisfy a subset of the service selection requirements of a user to a certain degree. Classical logics including description logic is ineffective to model such a satisfiability relation. Furthermore, classical logics are ineffective to model the non-monotonicity nature in web service selection and recommendation [15]. For instance, even if a user is interested in web services about "sports" today, it is quite possible for him/her to develop interests in other areas such as "investment" instead of "sports" at a later stage. This kind of changing service preferences (i.e., the non-monotonicity of service preferences) is quite common and expected to be supported by an effective service recommendation mechanism.

Though sophisticated quantitative methods have been develop to model the decision making processes [5, 13] and they may be applied to develop service recommendation agents, the main weakness of these quantitative approaches is that the axioms characterizing a particular decision theory may not be well-defined and it is not clear whether applying such a theory can lead to the maximal outcome given a particular situation [5]. Bayesian networks have also been explored for uncertainty analysis [10, 17]. However, it is difficult to obtain all the conditional probabilities required in a large network model. Though assuming independencies among events may simplify the calculations, the resulting model may not reflect the realities of the underlying problem domain. For service recommendation, it is desirable for the intelligent agents to explain and justify their decisions so that user trust and satisfaction can be enhanced [9, 15]. From this perspective, it is not easy to explain an agent's decisions purely based on a quantitative model where the relationships between various decision factors are buried in numerous conditional probabilities or weight vectors. A logic-based approach can offer the distinct advantage in terms of easily developing the explanations about intelligent agents' decision making behavior [23, 16].

The main contributions of this paper are three fold. First, a novel intelligent agent based web service recommendation and invocation system architecture is proposed. Second, a belief revision based service recommendation agent is developed to address the issues of non-monotonicity of service preferences. Third, through a rigorous quantitative evaluation under real-world web services recommendation and invocation scenarios, the proposed intelligent agent based web service recommendation mechanism is evaluated. The rest of the paper is structured as follows. Section 2 gives an overview of our proposed agent-based service recommendation and invocation system. Preliminaries of belief revision logic are provided in Section 3. The application of belief revision logic to knowledge representation and reasoning in service recommendation agents are discussed in Section 4. Section 5 illustrates the quantitative evaluation of the proposed service recommendation agents. Finally, future work is proposed, and a conclusion summarizing our findings is given.

2 An Overview of Intelligent Agent Based Service Recommendation

An overview of the proposed Agent-based Service Recommendation and Invocation (ASRI) framework is depicted in Figure 1. The Web site housing this system is certified digitally so that all the related parties can verify its identification. A user must first register with ASRI for autonomous service discover and recommendation. Users interact with their service recommendation agents who will in-turn looking up the most relevant web services from the external service registries on behalf of their users. Intelligent agents [12, 28] are computer software situated on a computing network environment (e.g., the Internet) for autonomous information processing on behalf of their users. The Query Processing and Logging module of ASRI is responsible for accepting users' service requests and managing the service request and invocation histories. The User Profiling module underpinned by belief revision logic is used to maintain a consistent user profile according to sound belief revision axioms. A user profile contains a set of beliefs characterizing the user's specific preferences for web services and their expectations about the functional requirements of the services. With reference to the request and service invocation history, the user profile can be revised to reflect the user's most recent interests.

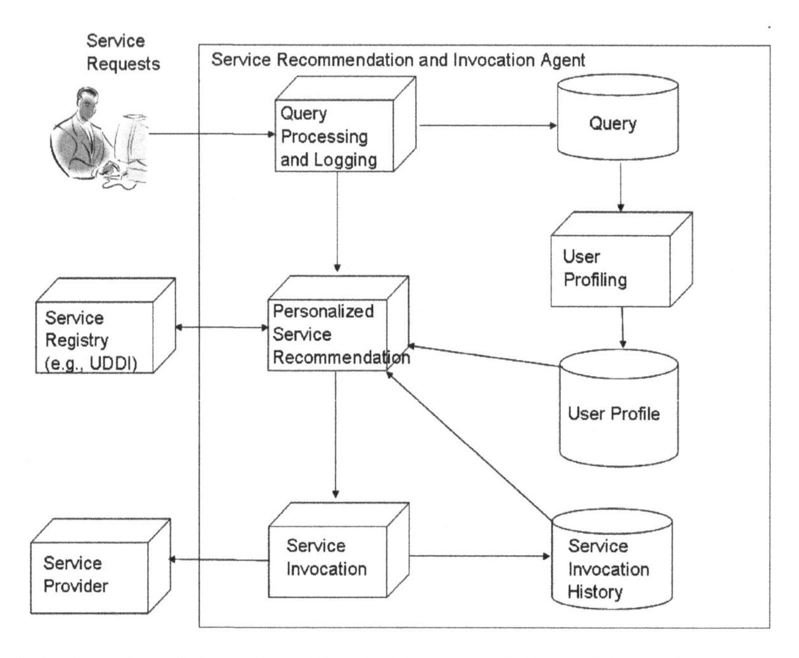

Fig. 1 An Overview of Agent-based Service Recommendation and Invocation

After a service request is accepted to ASRI, the Personalized Service Recommendation module will contextualize the original request by referring to the specific user profile. Belief revision based reasoning is applied to infer the most appropriate context for the initial service request with respect to the beliefs stored in the user's profile. After service request reasoning, the set of candidate web services are identified and ranked. The top ranked services are supposed to be the most desirable services matching the user's specific interests. Service invocation information about the ranked web services can be obtained from the external service registries after Web services looking-up. Finally, ASRI invokes the candidate services by communicating with the external service providers. Our prototype system was developed using Java JDK 6 and operated under Apache Tomcat 6.0. The belief revision engine used by our system is based on a Java implementation which has been successfully applied to adaptive information retrieval [23, 16].

3 Preliminaries of the AGM Belief Revision Logic

The AGM belief revision framework is coined after its founders Alchourrón, Gärdenfors, and Makinson [1]. It is one of the most influential works in the theory of belief revision. In this framework, belief revision processes are taken as the transitions among belief states. A belief state (set) K is represented by a theory of a classical language $\mathscr{L}$. A belief is represented by a sentence of $\mathscr{L}$ supplemented with an entrenchment degree indicating the degree of firmness of such a belief. Three principle types of belief state transitions are identified and modelled by the corresponding belief functions: *expansion* (K^{+}_{α}), *contraction* (K^{-}_{α}), and *revision* (K^{*}_{α}). The AGM framework comprises sets of postulates to characterise these functions for *consistent* and *minimal* belief revision. In addition, the AGM framework also specifies the constructions of the belief functions based on various mechanisms. One of them is *epistemic entrenchment* ($\leqslant$) [6]. It captures the notions of *significance*, *firmness*, or *defeasibility* of beliefs. If inconsistency arises after applying changes to a belief set, the least significant beliefs (i.e., beliefs with the lowest entrenchment degree) are given up in order to restore consistency. For a computer-based implementation of epistemic entrenchment and hence the AGM belief functions, Williams developed finite partial entrenchment rankings to represent epistemic entrenchment orderings [26].

Definition 1. A finite partial entrenchment ranking is a function $\mathbf{B}$ that maps a finite subset of sentences in $\mathscr{L}$ into the interval $[0,1]$ such that the following conditions are satisfied for all $\alpha \in dom(\mathbf{B})$:

(PER1) $\{\beta \in \text{dom}(\mathbf{B}) : \mathbf{B}(\alpha) < \mathbf{B}(\beta)\} \nvdash \alpha$;
(PER2) If $\vdash \neg\alpha$ then $\mathbf{B}(\alpha) = 0$;
(PER3) $\mathbf{B}(\alpha) = 1$ if and only if $\vdash \alpha$.

(PER1) states that the set of sentences ranked strictly higher than a sentence α cannot entail α. This property corresponds to the *Dominance* property of epistemic entrenchment [6]. $\mathbf{B}(\alpha)$ is referred to as the *degree of entrenchment* of an explicit belief α. The set of explicit beliefs of $\mathbf{B}$ is $\{\alpha \in dom(\mathbf{B}) : \mathbf{B}(\alpha) > 0\}$, and is denoted $exp(\mathbf{B})$. The set of implicit beliefs $K = \text{Cn}(exp(\mathbf{B}))$ is denoted $content(\mathbf{B})$, where Cn is the classical consequence operator.

In order to describe the epistemic entrenchment ordering ($\leqslant_{\mathbf{B}}$) generated from a finite partial entrenchment ranking $\mathbf{B}$, it is necessary to rank implicit beliefs.

Definition 2. Let $\alpha \in \mathscr{L}$ be a contingent sentence. Let $\mathbf{B}$ be a finite partial entrenchment ranking and $\beta \in exp(\mathbf{B})$. The degree of entrenchment of an implicit belief α is defined by:

$$degree(\mathbf{B},\alpha) = \begin{cases} sup(\{\mathbf{B}(\beta) \in ran(\mathbf{B}) : cut_{\leqslant}(\beta) \vdash \alpha\}) \\ \qquad \text{if } \alpha \in content(\mathbf{B}) \\ 0 \qquad \text{otherwise} \end{cases}$$

where the *sup* function returns the supremum from a set of ordinals. The $cut_{\leqslant}(\beta)$ operation extracts a set of explicit beliefs which is at least as entrenched as β according to an entrenchment ranking **B**. $\vdash$ is the classical inference relation. Precisely, a *cut* operation is defined by: $cut_{\leqslant}(\beta) = \{\gamma \in dom(\mathbf{B}) : \mathbf{B}(\beta) \leqslant \mathbf{B}(\gamma)\}$. For example, given the belief set $\mathbf{B} = \{$(grass-wet, 0.6), (grass-wet $\rightarrow$ rained, 0.5) $\}$, the operation $cut_{\leqslant}$(grass-wet) will return a single belief "grass-wet". Moreover, $degree(\mathbf{B}, \text{rained}) = 0.5$ is derived according to Definition 2 because the minimal entrenchment degree in the highest cut of **B** that logically entail ($\vdash$) "rained" is 0.5.

In a belief revision based service recommendation agent system, service requests and preferences are represented by some beliefs. When a service user's interests change, the entrenchment degrees of the corresponding beliefs are raised or lowered in the agent's knowledge base. Raising or lowering the entrenchment degree of a belief is conducted via a belief revision operation $\mathbf{B}^{\star}(\alpha, i)$ where α is a sentence and i is the new entrenchment degree. The Rapid Anytime Maxi-adjustment (RAM) method developed based on the Maxi-adjustment method [27] is proposed to implement the belief revision operator $\mathbf{B}^{\star}(\alpha, i)$:

Definition 3. Let α be a contingent sentence, $j = degree(\mathbf{B}, \alpha)$ and $0 \leqslant i < 1$. The (α, i) Rapid Anytime Maxi-adjustment of **B** is $\mathbf{B}^{\star}(\alpha, i)$ defined by:

$$\mathbf{B}^{\star}(\alpha, i) = \begin{cases} \mathbf{B}^{-}(\alpha, i) & \text{if } i < j \\ (\mathbf{B}^{-}(\neg\alpha, 0))^{+}(\alpha, i) & \text{if } i > j \\ \mathbf{B}^{+}(\alpha, i) & \text{if } i = j \text{ and } j > 0 \text{ and} \\ & \alpha \notin exp(\mathbf{B}) \\ \mathbf{B} & \text{otherwise} \end{cases}$$

where for all $\beta \in dom(\mathbf{B})$, $\mathbf{B}^{-}(\alpha, i)$ is defined as follows:

1. For β with $\mathbf{B}(\beta) > j$, $\mathbf{B}^{-}(\alpha, i)(\beta) = \mathbf{B}(\beta)$.
2. For β with $i < \mathbf{B}(\beta) \leqslant j$,

$$\mathbf{B}^{-}(\alpha, i)(\beta) = \begin{cases} i & \text{if } \{\gamma : \mathbf{B}^{-}(\alpha, i)(\gamma) > \mathbf{B}(\beta)\} \cup \\ & \{\delta : \mathbf{B}^{-}(\alpha, i)(\delta) = \mathbf{B}(\beta) \wedge \\ & Seq(\delta) \leq Seq(\beta)\} \vdash \alpha \\ \mathbf{B}(\beta) & \text{otherwise} \end{cases}$$

3. For β with $\mathbf{B}(\beta) \leqslant i$, $\mathbf{B}^{-}(\alpha, i)(\beta) = \mathbf{B}(\beta)$.

For all $\beta \in dom(\mathbf{B}) \cup \{\alpha\}$, $\mathbf{B}^{+}(\alpha, i)$ is defined as follows:

1. For β with $\mathbf{B}(\beta) \geqslant i$, $\mathbf{B}^{+}(\alpha, i)(\beta) = \mathbf{B}(\beta)$.
2. For β with $j \leqslant \mathbf{B}(\beta) < i$,

$$\mathbf{B}^{+}(\alpha, i)(\beta) = \begin{cases} i & \text{if } i < degree(\mathbf{B}, \alpha \rightarrow \beta) \\ degree(\mathbf{B}, \alpha \rightarrow \beta) & \text{otherwise} \end{cases}$$

3. For β with $\mathbf{B}(\beta) < j$, $\mathbf{B}^{+}(\alpha, i)(\beta) = \mathbf{B}(\beta)$.

The intuition of the RAM method is that if the new entrenchment degree i of a sentence α is less than its existing degree j, a contraction operation $\mathbf{B}^{-}(\alpha,i)$ (i.e., lowering its entrenchment degree) is invoked. If the new degree of α is higher than its existing degree, an expansion operation $\mathbf{B}^{+}(\alpha,i)$ will be initiated. Hence, $\neg\alpha$ must first be assigned the lowest entrenchment degree (i.e., contracting it from the theory). Then, the degree of α is raised to the new degree i. During raising or lowering the entrenchment degree of α, the degrees of other sentences are adjusted in a *minimal* way such that PER1, PER2, and PER3 are always maintained. The notation $\mathbf{B}^{-}(\alpha,i)(\beta)$ refers to the new entrenchment degree of β after applying the contraction operation (-) to the belief set $K = content(\mathbf{B})$. The *Seq* operator assigns an unique sequence number to a sentence if there is more than one sentence in a particular entrenchment rank. Under such a circumstance, it does not matter which sentence is contracted first because these sentences are equally preferred or not preferred from a rational agent's point of view. If a finite partial entrenchment ranking $\mathbf{B}$ has x natural partitions, it only requires $\log_2 x$ classical satisfiability checks [14]. Therefore, given the propositional Horn logic $\mathcal{L}_{Horn}$ as the representation language, the RAM method only involves polynomial time complexity.

4 Knowledge Representation and Reasoning

Based on a user's explicit and implicit feedback, the belief revision based user profiling module can induce a set of beliefs representing the user's current service needs. The user's changing preferences are then revised to the agent's knowledge base via the belief revision processes. The following are examples showing how the belief-based profile learning and service recommendation reasoning works.

4.1 Service Recommendation Situation SR_1

Table 1 summarizes the changes applied to the agent's knowledge base K before and after a service request is received. This example describes a service recommendation situation that a user is probably interested in service about "finance". Contextual knowledge is acquired via typical knowledge engineering processes (e.g., knowledge elicitation from the user) [15]. For instance, the entrenchment degree about "finance" is induced based on the frequency of the previous service invocation (e.g., 7 out of 10 times the user explicitly requesting that type of services). Previous work also employs a variant of Kullback-Leibler divergence to automatically induce the entrenchment degrees of beliefs [23, 16]. The upper half of Table 1 shows all the explicit beliefs and the lower half of the table lists some of the implicit beliefs. The before and the after columns show the entrenchment degrees of the beliefs before and after user profiling revision, respectively. For simplicity and easy of exposition, the examples are developed using the propositional language. However, the AGM belief revision logic can be implemented based on any classical languages such as predicate logic [26].

Table 1 The First Service Recommendation Situation SR_1

Beliefs (α)	Before	After
$\neg failure$	1.000	1.000
$finance \rightarrow investment$	0.800	0.800
$finance \rightarrow \neg art$	0.800	0.800
$cheap \wedge failure \rightarrow risky$	0.800	0.800
$finance$	0.000	**0.700**
$investment$	0.000	**0.700**
$\neg art$	0.000	**0.700**

To revise the user's current interest into the agent's knowledge base, Rapid Anytime Maxi-adjustment operation $\mathbf{B}^{\star}(finance, 0.7)$ is invoked. After such a learning process, the agent can automatically deduce that the user may be interested in services about "investment". By viewing the recommended services, the user may eventually find that she is interested or not interested in that service. The recommendation agent can then revise its beliefs about the user's preferences based on the user's relevance feedback or using an automated preference induction mechanism [23, 16]. The entrenchment degrees of the implicit beliefs (e.g., sentences listed in the lower half of the table) are computed according to Definition 2. Only the implicit beliefs relevant for our discussion are shown in the table. A belief with zero entrenchment degree is contracted from K.

4.2 Service Matching and Recommendation

An entrenchment-based similarity measure $Sim(SR,s)$ Eq.(1) is proposed to approximate the *semantic correspondence* between a service recommendation situation SR and a web service s which is represented by a service description d. A service recommendation situation refers to some service preferences and the associated contextual knowledge (e.g., $finance \rightarrow investment$).

$$\begin{aligned} Sim(SR,s) &\approx Sim(\mathbf{B},d) \\ &= \frac{\sum_{l\in d}[degree(\mathbf{B},l)-degree(\mathbf{B},\neg l)]}{|L|} \end{aligned} \tag{1}$$

Eq.(1) combines the advantages of quantitative ranking and symbolic reasoning in a single formulation. It is not a simple overlapping model since the function $degree(\mathbf{B},l)$ invokes non-monotonic reasoning about the relevance of a service description d with respect to the knowledge base $content(\mathbf{B})$ which represents a service recommendation situation SR. The basic idea is that a service description d is characterized by a set of positive literals (i.e., service characteristics) $d = \{l_1, l_2, \ldots, l_n\}$. If the agent's knowledge base $K = content(\mathbf{B})$ logically entails an atom l_i, a positive contribution is made to the overall similarity score because of the partial semantic correspondence between SR and d. This kind of logical entailment is non-classical and is implemented based on the function *degree* defined in Definition 2. Conceptually, service matching is underpinned by $SR \mathrel{\mid\!\sim_K} d$, where $\mid\!\sim_K$ is a non-monotonic inference relation [7]. On the other hand, if K implies the negation of a literal $l_i \in d$, it shows the *semantic distance* between SR and d. Therefore, the similarity value is reduced by a certain degree. The set L is defined by $L = \{l \in d : degree(\mathbf{B}, l) > 0 \vee degree(\mathbf{B}, \neg l) > 0\}$. As an example, if three web services are evaluated by the recommendation agent with reference to the first service recommendation situation SR_1, the services are ranked as follows:

$$d_1 = \{finance, \neg failure\}$$
$$d_2 = \{investment, failure\}$$
$$d_3 = \{art, \neg failure\}$$

$$\because Sim(\mathbf{B}, d_1) = 0.85$$
$$Sim(\mathbf{B}, d_2) = -0.15$$
$$Sim(\mathbf{B}, d_3) = 0.15$$

$$\therefore doc_2 \preceq doc_3 \preceq doc_1$$

where $d_i \preceq d_j$ means d_j is at least as preferable as d_i with respect to a service recommendation situation. Such a ranking corresponds to our intuition about web service preference for the situation SR_1.

4.3 Learning in Retrieval Situation SR_2

If the service recommendation context is changed because the user is interested in web services about "art" later on, the agent's knowledge base K before and after incorporating such a contextual change is depicted in Table 2. The new information about the user's preferences is revised into K via the belief revision operation $\mathbf{B}^{\star}(art, 0.9)$. In this case, the entrenchment degree of the belief is assumed to be induced automatically or via direct user feedback. According to the RAM method, $((\mathbf{B}^{-}(\neg art, 0))^{+}(art), 0.9)$ is executed. As the belief $(\neg art, 0.700)$ is contained in

Table 2 The Second Service Recommendation Situation SR_2

Beliefs (α)	Before	After
$\neg failure$	1.000	1.000
art	0.000	**0.900**
$finance \rightarrow investment$	0.800	0.800
$cheap \wedge failure \rightarrow risky$	0.800	0.800
$finance \rightarrow \neg art$	0.800	**0.000**
$finance$	0.700	**0.000**
$investment$	0.700	**0.000**
$\neg art$	0.700	**0.000**

K, the belief revision operation must first lower the entrenchment degree of $\neg art$ to zero before adding the explicit belief $(art, 0.9)$ into K such that the representation of the service recommendation situation remains consistent and coherent. In doing so, the explicit belief $(finance \rightarrow \neg art, 0.800)$ (i.e., the least entrenched belief) is contracted from the theory base $exp(\mathbf{B})$. If a dispatch threshold t is used, the agent can make binary decisions for service recommendation. For instance, if $t = 0.55$ is chosen, the agent will select d_1 for the user in the first recommendation situation, but select d_3 in the second situation. The service ranking in recommendation situation SR_2 is as follows:

$$\because Sim(\mathbf{B}, d_1) = 0.500$$
$$Sim(\mathbf{B}, d_2) = -0.500$$
$$Sim(\mathbf{B}, d_3) = 0.950$$

$$\therefore doc_2 \preceq doc_1 \preceq doc_3$$

5 Experiments and Results

To evaluate the effectiveness of the proposed belief revision based service recommendation agents, we first constructed a set of user profiles characterizing the service interests and QoS requirements for a wide range of domains such as sport, finance, weather, car, insurance, and shipping. For each user profile, a set of 50 relevant web services is identified by two human annotators. These web services are first identified via Web sites (e.g., xmethods.net, webservicex.net) which show publicly available web services and web service search engines (e.g., webservices.seekda.com, esynaps.com). A total of 1,000 web services were identified and used in this experiment. Figure 2 is a snapshot view of the inquired web services via the seekda web service search engine.

We used the WSDL2 Java tool from the Axis2 package to develop the client stub classes to invoke the respective web services, and hence to collect the corresponding QoS attributes. The description d of each web service s was composed by two human annotators; the QoS attributes as well as the keywords corresponding to the service which was made available via web service search engines were included in the service description d. Changing web service recommendation scenarios were simulated by manually revising the user profiles such as changing a profile from "sport" to "insurance", or from "insurance" to "weather", etc. Each user profile was injected the changes once, and the precision, recall and F-measure [20] of the web services recommended by the agent were then computed. The precision, recall, and F-measure are defined by the following formulas whereas a, b, c, d represent the number correctly recommended web services, the number of incorrectly recommended web services, the number of not recommended relevant web services, and the number of not recommended non-relevant web services respectively.

$$\text{Precision} = \frac{a}{a+b} \tag{2}$$

$$\text{Recall} = \frac{a}{a+c} \tag{3}$$

$$\text{F-measure} = \frac{2a}{2a+b+c} \tag{4}$$

After the changes of service preferences were introduced to a user profile, the service recommendation agent should exercise non-monotonic inference to re-rank the potential web services, and hopefully with the relevant web services (pre-defined by

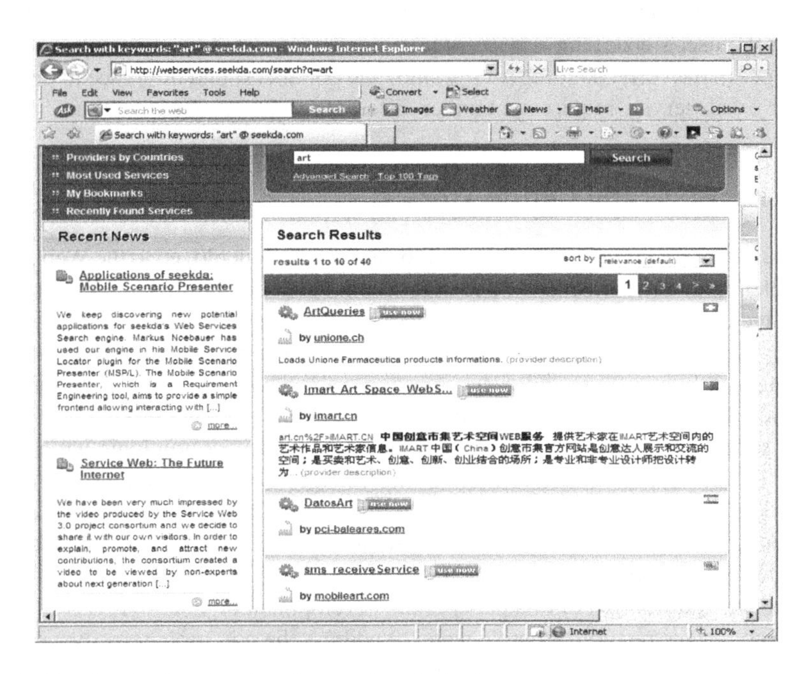

Fig. 2 Web Service Inquiry via a Service Search Engine

Table 3 Comparative Performance of Service Recommendation Agent

Service	Before		After		
Category	Recall	Precision	Recall	Precision	Δ F-measure
Sport	0.652	0.748	0.650	0.742	-0.52%
Finance	0.612	0.674	0.608	0.665	-0.97%
Weather	0.610	0.687	0.602	0.673	-1.65%
Car	0.664	0.733	0.662	0.726	-0.62%
Insurance	0.606	0.713	0.604	0.701	-0.98%
Shipping	0.608	0.707	0.600	0.696	-1.44%
Average	0.625	0.710	0.620	0.694	-1.58%

human annotators) ranked at the top of the recommendation list. The preliminary experimental result about the effectiveness of the service recommendation agent is depicted in Table 3. The predictions of the service recommendation agent achieve 0.625 (recall) and 0.710 (precision) on average before any functional or preferential changes occur. As can be observed, the changes of precision and recall figures are minimal after the preferential changes are injected to the respective user profiles. The F-measure is only dropped by -1.58% on average with the presence of changes related to functional or preferential service requirements. The reason is that the proposed belief revision logic is effective to model the changes of beliefs within an agent's knowledge base. Essentially, the service recommendation agent can keep track of the user's changing service requirements (simulated by artificial injection of changes to the corresponding user profile) and conduct sound inference to predict the user's web service requirement under different service request situations.

6 Conclusions and Future Work

With the explosive growth of the information and services deployed to the Web these days, intelligent and personalized services recommendation becomes desirable. Although recent research in semantic web services has highlighted the advantages of applying formal knowledge representation and reasoning models to improve the effectiveness in human-to-machine and machine-to-machine interactions, the issues of service non-monotonicity and gradated logical inference have received little attention so far. The problem of classical logics such as description logic which underpins the development of OWL is that the changes of service preferences in an agent's knowledge base cannot be effectively modeled. Moreover, classical logics are often restricted by the binary inference relation which cannot effectively handle a gradated assessment of web service quality or preferential requirements.

The main contribution of this paper is the development of a belief revision logic based web service recommendation agent to address the non-monotonicity issue of web service recommendation. Our preliminary experiment conducted based on real-world web service recommendation scenarios shows that the proposed logical model is effective for service recommendation. The service recommendation agents underpinned by the proposed belief revision logic are able to keep track of users' changing service requirements and maintain the effectiveness of service recommendation over time. To the best of our knowledge, our work is the first successful application of belief revision logic to model the non-monotonicity in service recommendation agents. Future work involves comparing the effectiveness of the proposed logic-based service recommendation agents and that of other quantitative approaches of service recommendation under a variety of real-world web service discovery and invocation scenarios. The scalability of the belief revision based service recommendation agents will be examined by using a larger scale experiment.

References

1. Alchourrón, C.E., Gärdenfors, P., Makinson, D.: On the logic of theory change: partial meet contraction and revision functions. Journal of Symbolic Logic 50, 510–530 (1985)
2. Berners-Lee, T., Hendler, J., Lassila, O.: The semantic web. Scientific American 284(5), 34–43 (2001)
3. The World Wide Web Consortium. Web Ontology Language (2004), `http://www.w3.org/2004/OWL/`
4. Diaz, O.G.F., Salgado, R.S., Moreno, I.S.: Using case-based reasoning for improving precision and recall in web services selection. International Journal of Web and Grid Services 2, 306–330 (2006)
5. Edwards, W.: Utility Theories: Measurements and Applications. Kluwer Academic Publishers, Norwell (1992)
6. Gärdenfors, P., Makinson, D.: Revisions of knowledge systems using epistemic entrenchment. In: Vardi, M.Y. (ed.) Proceedings of the Second Conference on Theoretical Aspects of Reasoning About Knowledge, pp. 83–95. Morgan Kaufmann, Pacific Grove (March 1988)
7. Gärdenfors, P., Makinson, D.: Nonmonotonic inference based on expectations. Artificial Intelligence 65(2), 197–245 (1994)
8. Guo, W.Y.: Reasoning with semantic web technologies in ubiquitous computing environment. Journal of Software 3(8), 27–33 (2008)
9. He, M., Jennings, N.R., Leung, H.: On agent-mediated electronic commerce. IEEE Trans. on Knowledge and Data Engineering 15(4), 985–1003 (2003)
10. Hrycej, T.: Gibbs sampling in bayesian networks. Artificial Intelligence 46(3), 351–363 (1990)
11. Huang, Z., Lu, X., Duan, H.: Context-aware recommendation using rough set model and collaborative filtering. Artificial Intelligence Review 35(1), 85–99 (2011)
12. Jennings, N., Sycara, K., Wooldridge, M.: A roadmap of agent research and development. Journal of Autonomous Agents and Multi-Agent Systems 1(1), 7–38 (1998)
13. Keeney, R., Raiffa, H.: Decisions with Multiple Objectives: Preferences and Value Tradeoffs. Cambridge University Press, Cambridge (1993)
14. Lang, J.: Possibilistic Logic: Algorithms and Complexity. In: Kohlas, J., Moral, S. (eds.) Handbook of Algorithms for Uncertainty and Defeasible Reasoning. Kluwer Academic Publishers, Norwell (1997)
15. Lau, R.Y.K., Zhang, W.: Non-monotonic modeling for personalized services retrieval and selection. International Journal of Systems and Service-Oriented Engineering 1(2), 55–68 (2010)
16. Lau, R.Y.K., Bruza, P., Song, D.: Towards a Belief Revision Based Adaptive and Context-Sensitive Information Retrieval System. ACM Transactions on Information Systems 8, 1–8 (2008)
17. Lau, R.Y.K., Wong, O., Li, Y., Ma, L.C.K.: Mining Trading Partners' Preferences for Efficient Multi-Issue Bargaining in e-Business. Journal of Management Information Systems 25(1), 81–106 (2008)
18. Liu, H., Li, Q., Gu, N., Liu, A.: A logical framework for modeling and reasoning about semantic web services contract. In: Huai, J., Chen, R., Hon, H.-W., Liu, Y., Ma, W.-Y., Tomkins, A., Zhang, X. (eds.) Proceedings of the 17th International Conference on World Wide Web, WWW 2008, Beijing, China, April 21-25, pp. 1057–1058. ACM (2008)

19. Roman, D., Kifer, M.: Reasoning about the behavior of semantic web services with concurrent transaction logic. In: Koch, C., Gehrke, J., Garofalakis, M.N., Srivastava, D., Aberer, K., Deshpande, A., Florescu, D., Chan, C.Y., Ganti, V., Kanne, C.-C., Klas, W., Neuhold, E.J. (eds.) Proceedings of the 33rd International Conference on Very Large Data Bases, University of Vienna, Austria, September 23-27, pp. 627–638. ACM (2007)
20. Salton, G., McGill, M.J.: Introduction to Modern Information Retrieval. McGraw-Hill, New York (1983)
21. Senivongse, T., Phacharintanakul, N., Ngamnitiporn, C., Tangtrongchit, M.: A capability granularity analysis on web service invocations. In: Ao, S.I., Douglas, C., Grundfest, W.S., Burgstone, J. (eds.) Proceedings of the World Congress on Engineering and Computer Science. Lecture Notes in Engineering and Computer Science, vol. I, pp. 400–405. Newswood Limited (2010)
22. Skonnard, A.: Publishing and Discovering Web Services with DISCO and UDDI (February 2002), http://msdn.microsoft.com/msdnmag/
23. Song, D., Lau, R.Y.K., Bruza, P.D., Wong, K.F., Chen, D.Y.: An adaptive information agent for document title classification and filtering in document-intensive domains. Decision Support Systems 44(1), 251–265 (2008)
24. Steller, L., Krishnaswamy, S.: Efficient mobile reasoning for pervasive discovery. In: Shin, S.Y., Ossowski, S. (eds.) Proceedings of the 2009 ACM Symposium on Applied Computing, SAC, Honolulu, Hawaii, USA, March 9-12, pp. 1247–1251. ACM (2009)
25. Varga, L.Z., Hajnal, Á.: Engineering Web Service Invocations from Agent Systems. In: Mařík, V., Müller, J.P., Pěchouček, M. (eds.) CEEMAS 2003. LNCS (LNAI), vol. 2691, pp. 626–635. Springer, Heidelberg (2003)
26. Williams, M.-A.: Iterated theory base change: A computational model. In: Mellish, C.S. (ed.) Proceedings of the Fourteenth International Joint Conference on Artificial Intelligence, August 20–25, pp. 1541–1547. Morgan Kaufmann Publishers, Montréal (1995)
27. Williams, M.-A.: Anytime belief revision. In: Pollack, M.E. (ed.) Proceedings of the Fifteenth International Joint Conference on Artificial Intelligence, Nagoya, Japan, August 23–29, pp. 74–79. Morgan Kaufmann (1997)
28. Wooldridge, M., Jennings, N.: Intelligent Agents: Theory and Practice. Knowledge Engineering Review 10(2), 115–152 (1995)
29. Yang, S.J.H., Zhang, J., Lan, B.C.W.: Service-level agreement-based qoS analysis for web services discovery and composition. Int. J. of Internet and Enterprise Management 5, 39–58 (2006)

Single Character Frequency-Based Exclusive Signature Matching Scheme

Yuxin Meng, Wenjuan Li, and Lam-for Kwok

Abstract. Currently, signature-based network intrusion detection systems (NIDSs) have been widely deployed in various organizations such as universities and companies aiming to identify and detect all kinds of network attacks. However, the big suffering problem is that signature matching in these detection systems is too expensive to their performance in which the cost is at least linear to the size of an input string and the CPU occupancy rate can reach more than 80 percent in the worst case. This problem is a key limiting factor to encumber higher performance of a signature-based NIDS under a large-scale network. In this paper, we developed an exclusive signature matching scheme based on single character frequency to improve the efficiency of traditional signature matching. In particular, our scheme calculates the single character frequency from both stored and matched NIDS signatures. In terms of a decision algorithm, our scheme can adaptively choose the most appropriate character for conducting the exclusive signature matching in distinct network contexts. In the experiment, we implemented our scheme in a constructed network environment and the experimental results show that our scheme offers overall improvements in signature matching.

Keywords: Intrusion detection, Signature matching, Intelligent system, Network security.

1 Introduction

Network threats (e.g., virus, worm, trojan) have become a big challenge with regard to current network communications. Thus, network intrusion detection systems

Yuxin Meng · Lam-for Kwok
Department of Computer Science, City University of Hong Kong, Hong Kong SAR, China
e-mail: ymeng8@student.cityu.edu.hk, cslfkwok@cityu.edu.hk

Wenjuan Li
Computer Science Division, Zhaoqing Foreign Language College, Guangdong, China
e-mail: wenjuan.anastatia@gmail.com

R. Lee (Ed.): Computer and Information Science 2012, SCI 429, pp. 67–80.
springerlink.com

(NIDSs) [1, 3] have been widely implemented in a variety of organizations with the purpose of defending against different kinds of network attacks. Take an assurance company for an example, network security is the key element to the whole organization since a very small security error may cause a loss of millions of dollars. To enhance the network security, NIDSs have become an essential component to the security infrastructure of these organizations.

In general, the network intrusion detection systems can be classified into two categories[1]: *signature-based NIDS* and *anomaly-based NIDS*. As the name suggests, the signature-based NIDS [5] (also called *misuse-based NIDS* or *rule-based NIDS*) detects an attack by comparing its signatures[2] with network packet payloads. In this case, such kind of NIDSs can only detect well-known attacks. While the anomaly-based NIDS [7, 8] can identify novel attacks in that it detects an attack by discovering great deviations between its established normal profile and observed events. A normal profile is used to represent the normal behavior of a user or network. In fact, the signature-based NIDS is much more popular than the anomaly-based NIDS in the real deployment among various organizations since the number of false alarms of the signature-based NIDS is far less than that of the anomaly-based NIDS [6]. For instance, a company could receive hundreds of alarms per day by using a signature-based NIDS, but it maybe obtain more than thousands of alarms per day by means of an anomaly-based NIDS.

Problem. Although the signature-based network intrusion detection systems are epidemic in real settings, their poor performance in the high volume traffic environment is still a big challenge in encumbering their further development. For example, Snort [3, 9] which is a light-weight signature-based NIDS can quickly exhaust a computer memory especially in the high volume traffic environment [4]. In this case, Snort has to drop huge amounts of network packets which will cause a lot of security issues (i.e., missing some deleterious packets). The major consuming time of a NIDS is spent in comparing their signatures with incoming network packets in which the computing consumption is at least linear to the size of an input string [10]. Take Snort for an example, it spends about 30 percent of its total processing time in comparing the incoming packet payloads with its signatures while the consuming time even can reach more than 80 percent when deployed in the intensive web-traffic environment [11].

Contributions. To mitigate the above problem, we advocate that designing an efficient signature matching scheme is a promising way to improve the performance of a signature-based NIDS under the high volume traffic environment. In this paper, we propose and develop an exclusive signature matching scheme based on single character frequency intended to improve the performance of signature matching in NIDSs. In addition, our scheme can be adaptive in selecting the most appropriate single character for exclusive signature matching in terms of different network environments. In particular, our scheme respectively calculates the single character frequency of both stored NIDS signatures and matched signatures so as to adap-

[1] Another NIDS type is *stateful protocol analysis* [2] which adds stateful characteristics to regular protocol analysis.

[2] The *signature* is a kind of pattern to describe a well-known attack.

tively and sequentially determine the most appropriate single character for comparing with packet payload. Moreover, we implemented our exclusive signature matching scheme in a constructed network environment to evaluate its performance. The initial experimental results show that our exclusive signature matching scheme can improve the original exclusion-based signature matching algorithm in the aspect of implementation (see Section 3) and can greatly reduce the consuming time in signature matching with respect to the traditional signature matching, which overall enhance the performance of a signature-based NIDS.

Motivation. The heavy burden of signature matching is a key limiting factor to impede higher performance of a NIDS in a larger-scale network environment. Although a lot of research efforts have been made aiming to address this problem (see Section 2), this issue is still a hot topic. Many commercial companies greatly expect more efficient signature matching schemes to level up their NIDSs for the sake of improving the performance of their networks in the aspect of security. In such a requirement, we make an effort to further develop an intelligent and efficient signature matching scheme based on character frequency under intensive web-traffic environments.

The remaining parts of this paper are organized as follows: in Section 2, we introduce background information of the original exclusion-based signature matching and describe some related work about the methods of improving signature matching process; we illustrate the architecture of our developed single character frequency-based exclusive signature matching scheme in Section 3 and we describe our experimental methodology and experimental results in Section 4; Section 5 states our future work; at last, we conclude our work in Section 6.

2 Background

In this section, we begin by briefly introducing the research work of exclusion-based signature matching algorithm and then we describe some related work about improving the performance of signature matching including the approaches of designing string matching algorithms, packet classification and packet filtration.

2.1 Exclusion-Based Signature Matching Algorithm

The exclusion-based signature matching algorithm (called *ExB*) was first proposed by Markatos *et al.* [12] that is a multiple-string matching algorithm designed for NIDSs. The basic idea is to determine if an input string contains all fixed-size bit-strings of a signature string without considering the bit-string sequence. For each incoming packet, *ExB* first creates an *occurrence bitmap* to guarantee that each fixed-size bit-string exists in the packet. Then, the bit-strings for each signature are matched against the occurrence bitmap. Soon afterwards, they presented an algorithm of E^2xB [13] as an improvement for the *ExB* in better using cells and supporting case-sensitive matching. In general, *ExB* and E^2xB are both based on the simple observation [12, 13] as below.

- Observation. Suppose there are two strings: an input string I and a small string s ($|s| < |I|$)[3]. The task is to check whether string I contains small string s. The observation is that if there exists at least one character of string s that is not contained in I, then s is not a sub-string of I.

In other words, if string s contains at least one character that is not in string I, then string s is not a sub-string of string I. This simple observation can be used to quickly determine a given string s does not appear in the input string I. While if every character of string s is contained in the input string I, then the exclusion-based signature matching algorithm still need a standard string matching algorithm to confirm whether string s is actually a sub-string of I. In this case, this method is effective and efficient when there is a fast way of checking whether a given string s belongs to I or not.

2.2 Related Work

For string matching algorithm, the fast string searching algorithm designed by Boyer and Moore [14] is the most widely used single pattern matching algorithm which uses two heuristics to cut down the number of searches during the matching process and begins with the rightmost character of a target string. Horspool [15] then improved Boyer-Moore algorithm by using only the bad character heuristic to achieve a better matching speed. For multi-pattern matching algorithm, the Aho-Corasick algorithm [17] is well known in organizing patterns to construct a deterministic finite automaton (DFA) and searching all strings at the same time. The Wu-Manber algorithm [18], as an improvement for the Aho-Corasick algorithm, is to use no less than two hash tables based on the bad character heuristic to search multi patterns during the matching process. Several other improvements with regard to the well-known Aho-Corasick algorithm can be referred to work [19, 20].

It is worth mentioning that Fisk and Varghese [11] first considered to design a NIDS-specific string matching algorithm which they called *Set-wise Boyer-Moore-Horspool*. Their work showed that their algorithm is faster than the algorithms of Aho-Corasick and Boyer-Moor regarding to medium-size pattern sets.

Packet classification is another way that can efficiently improve the performance of signature matching. Lakshman and Stidialis [21] presented a packet classification scheme by using a worst case and traffic-independent performance metric to check among a few thousand filtering rules. That is, their method used linear time (or used a linear amount of parallelism) to sequentially search through all rules. Song and Lockwood [22] presented a novel packet classification framework called *BV-TCAM* by using FPGA technology for a FPGA-based Network Intrusion Detection System (NIDS) and the classifier can report multiple matches at gigabit per second network link rates. While Baboescu and Varghese [16] argued that hardware solutions do not scale to large classifier, and proposed two new ideas: recursive aggregation of bit maps and filter rearrangement. They used the ideas to design an algorithm called *Aggregated Bit Vector* (ABV) through improving bit vector search algorithm (BV).

[3] $|s|$ means the length of string s and $|I|$ means the length of string I.

The previous work [23, 24] has shown that the problem of packet classification is inherently hard.

In addition, packet filtration technique is an important and promising way of improving the performance of a NIDS by reducing the burden of signature matching in the high volume traffic environment. Meng and Kwok [25] proposed a novel method of using blacklist to construct an adaptive network packet filter in reducing the number of network packets that are required to be processed by a NIDS. They initially used a statistic-based blacklist generation method to obtain a blacklist in terms of IP confidence and showed encouraging results in their experiments.

3 Our Proposed Method

In this section, we begin with briefly introducing *Snort rules* that were used in our experiment to evaluate the performance of our scheme. Then, we give an in-depth description of our single character frequency-based exclusive signature matching scheme. Finally, we present a running example to illustrate the exclusive signature matching in detail.

3.1 Snort Rule

Snort [3, 9] is an open-source signature-based network intrusion detection system which has been widely used in the research area of intrusion detection. It maintains a rule (also called *signature*) database to detect various network attacks by means of signature matching and keep an update to the database periodically.

To facilitate the understanding of Snort rules, we give the generic format of Snort rule in Fig. 1 (a), which contains 8 parts: *action type*, *protocol type*, *source IP address*, *source port number*, *destination IP address*, *destination port number*, *content* and *message (msg)*. In Fig. 1 (b), we provide two specific examples of Snort rules. The first one aims to detect a kind of SNMP misc-attack with the signature '|04 01 00|'. While the second rule tries to identify an attack of webtrends scanner by using the signature '$|0A|help|0A|quite|0A|$'. The target packet of these two rules is UDP packet. $EXTERNAL_NET represents the source IP address and $HOME_NET represents the destination IP address.

When a packet arrives, Snort will first examine its IP address and port number to choose corresponding rules. Then, Snort will compare the packet payload with its selected one or several signatures. The commonly used two signature matching algorithms are Boyer-Moore [14] and Aho-Corasick [17]. Finally, Snort will produce an alarm if it successfully finds a match.

3.2 Our Proposal

To further improve the signature matching in NIDSs, we propose and develop a scheme of *single character frequency-based exclusive signature matching*.

Action-type protocol-type Source-ip Source-port -> Destination-ip Destination-port (content:"|attack signature|"; msg:"attack msg";)

(a) Generic rule format in Snort

alert udp $EXTERNAL_NET any -> $HOME_NET 161 (msg:"SNMP null community string attempt"; content:"|04 01 00|"; reference:cve,1999-0517; classtype: misc-attack; sid:1892; rev:8;)

alert udp $EXTERNAL_NET any -> $HOME_NET any (msg:"SCAN Webtrends Scanner UDP Probe"; flow:to_server; content: "|0A|help|0A|quite|0A|"; classtype: attempted-recon; sid:637; rev:6;)

(b) Two examples of Snort rules

Fig. 1 (a) Generic Snort rule format and (b) Two specific examples of Snort rules.

Our scheme refers to the basic idea of exclusion-based signature matching algorithm that identifying a mis-match rather than confirming an accurate match in the signature matching. Differently, our scheme uses an easier and intelligent implementation approach, and we investigate the effect of single character frequency on this kind of signature matching which had not been considered in the original exclusion-based signature matching algorithm. Therefore, we propose and use the term of *exclusive signature matching* instead of exclusion-based signature matching to distinguish our scheme from previous research work.

In addition to the simple observation in the original exclusion-based signature matching, our scheme is based on another observation that most packet payloads will not match any NIDS signature in most cases [26]. The observation which our scheme relies on has been verified in various experiments. We illustrate the architecture of our scheme in Fig. 2.

As shown in Fig. 2, our proposed scheme mainly contains four tables: a *table of stored NIDS signatures (SNS)*, a *table of matched NIDS signatures (MNS)* and two tables of *single character frequency (SCQ1 and SCQ2)*. The *table of stored NIDS signatures (SNS)* contains all active NIDS signatures that are stored in its rule database. While the *table of matched NIDS signatures (MNS)* contains all NIDS signatures which have ever been matched during the detection. The *table of SCQ1* aims to count the single character frequency according to the *table of SNS* and the *table of SCQ2* is responsible for calculating the single character frequency in terms of the *table of MNS*. In this case, we can obtain two tables with respect to the frequency of single characters. The single characters in these two *SCQ* tables are both arranged in descending order from most frequency to less frequency.

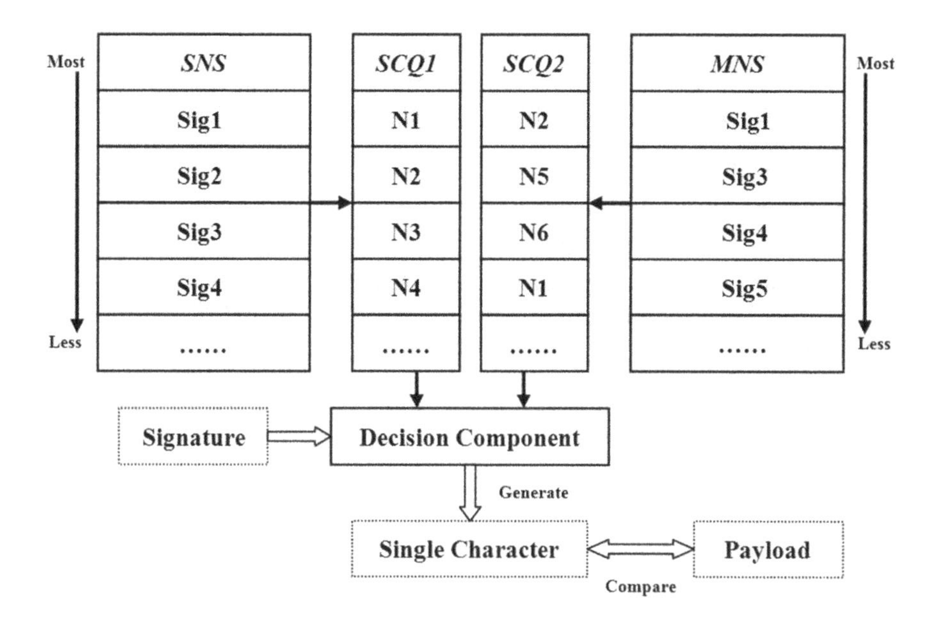

Fig. 2 The architecture of single character frequency-based exclusive signature matching scheme.

Decision component is used to determine and output the most appropriate single character from the above two *SCQ* tables for comparing with incoming packet payloads. To improve the effectiveness of our scheme, we give higher priority to the single character in the *table of SCQ2* since the *SCQ2* can be adaptive to specific context (i.e., *SCQ2* is responsible for counting the single character frequency based on matched NIDS signatures). In particular, the NIDS signatures in the *table of MNS* are used more frequent than the signatures in the *table of SNS*. Therefore, the statistical data in the *table of SCQ2* is more accurate (with regard to the *character frequency*) than the *table of SCQ1* as some (or even *most*) signatures in the *table of SNS* will have not matched any packet payload in a specific environment. This is the major reason that we give higher priority to the *table of SCQ2*. The detailed algorithm (also called *decision algorithm*) implemented in the decision component is described as below:

1. Input a signature s and a packet payload l; Set round number $r = 1$.
2. For each single character $a_i \in s$ $(i \in N, 1 \leq i \leq |s|)$. Finding the least frequent single character a_r (from single characters which have not been output) in the table *SCQ2*.

   - If a_r does not exist, then go to step 4,
   - If a_r exists, then go to step 3.

3. The decision component outputs a_r. Checking whether a_r is contained in l.

- If a_r is not contained in l, then the exclusive signature matching is completed. Signature s is not contained in packet payload l.
- If a_r is contained in l and $r < |s|$, then $r = r+1$, return to step 2, else if a_r is contained in l and $r == |s|$, then go to step 5.

4. Finding the least frequent single character a_r in the table *SCQ1*. The decision component outputs a_r. Checking whether a_r is contained in l.
   - If a_r is not contained in l, then the exclusive signature matching is completed. Signature s is not contained in packet payload l.
   - If a_r is contained in l and $r < |s|$, then $r = r+1$, return to 2, else if a_r is contained in l and $r == |s|$, then go to step 5.
5. All single characters in signature s are contained in packet payload l. Then using a traditional signature matching algorithm (e.g., Aho-Corasick) to check whether it is true or not.

3.3 Running Example

In order to make our scheme more easily understood, we give a light-weight running example to illustrate our *single character frequency-based exclusive signature matching* in detail. We randomly select three Snort rules from its *scan.rules* file (which is stored in Snort rule database) as below.

- alert tcp $EXTERNAL_NET any – > $HOME_NET 113 (msg: "SCAN ident version request"; content: "VERSION|0A|"; reference: arachnids, 303; classtype: attempted-recon; sid: 616; rev: 4;)
- alert udp $EXTERNAL_NET any – > $HOME_NET 10080:10081 (msg: "SCAN Amanda client-version request"; content: "Amanda"; classtype: attempted-recon; sid: 634; rev: 5;)
- alert udp $EXTERNAL_NET any – > $HOME_NET 7 (msg: "SCAN cybercop udp bomb"; content: "cybercop"; reference: arachnids, 363; classtype: bad-unknown; sid: 636; rev: 3;)

In addition to the above three rules, we further assume that only the signature "Amanda" has been matched. Therefore, we can construct the relevant four tables (*SNS*, *SCQ1*, *SCQ2* and *MNS*) as shown in Fig. 3. In particular, there are three signatures in *SNS* but only one signature in *MNS* (which is based on our assumption). The tables of *SCQ1* and *SCQ2* calculate the single character frequency in terms of *SNS* and *MNS* respectively. Based on our selected Snort rules, there are five characters in *SCQ2* while up to twenty characters in *SCQ1*. The single characters in both *SCQ1* and *SCQ2* are organized in columns from left to right based on frequency.

For the exclusive signature matching, we assume that an incoming packet with its payload string "Ameidetn". If the signature "Amanda" is selected for the signature comparison. Our scheme will sequentially check the single character '*n*', '*m*', '*d*', '*A*', '*a*' according to *SCQ2* and we can find that the single character '*a*' is not

SNS	*SCQ1*	*SCQ1*	*SCQ2*	*MNS*
VERSION|0A|	**a**	**y**	**a**	**Amanda**
Amanda	**A**	**E**	**A**	
cybercop	**c**	**I**	**d**	
	b	**N**	**m**	
	e	**O**	**n**	
	m	**R**		
	n	**S**		
	o	**V**		
	p	**0**		
	r	**|**		

Fig. 3 Four major tables of *SNS*, *SCQ1*, *SCQ2* and *MNS* in the running example.

contained in the payload string. Therefore, our exclusive signature matching is completed since the payload "Ameidetn" does not contain character 'a' that is contained in the signature "Amanda".

For another example, if the incoming payload string is "VERSION" and selected signature is "VERSION|0A|". In this case, our scheme can find that *'A'* is contained in this payload based on *SCQ2* but cannot further find other single characters in *SCQ2*. Therefore, our scheme, in terms of the *decision algorithm*, will search within the table of *SCQ1* and compare the single character '|' with the payload string. Fortunately, our scheme can quickly find that the single character '|' is not contained in the payload so that the exclusive signature matching is completed. The signature and packet payload are not matched.

4 Evaluation

In this section, we construct an experimental environment to evaluate the initial performance of our *single character frequency-based exclusive signature matching scheme*. The environmental deployment is presented in Fig. 4.

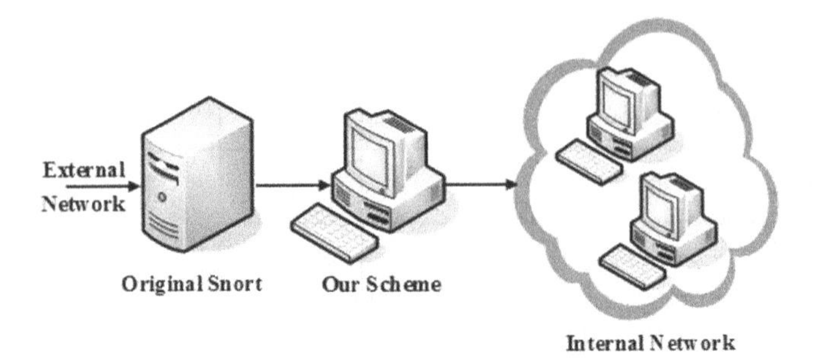

Fig. 4 Experimental environment with Snort and our scheme.

4.1 Experimental Methodology

As illustrated in Fig. 4, we deployed both the original Snort (version 2.9.0.5) and our scheme in frond of an internal network (which consists of a sever and several VM hosts). We implemented the default Snort rule set in both the Snort and our scheme. The internal network can communicate with the external network (Internet), therefore, all packets come from the external network to the internal network have to pass through both the Snort and our scheme. In this case, we could evaluate the performance of our scheme by comparing with the performance of Snort in terms of completion time[4] *versus* number of rules.

4.2 Evaluation Results

Following the experimental methodology, we launched our experiment and recorded related experimental data for two weeks. As illustrated in Fig. 5, we show the performance of our scheme compared to the performance of Snort in the experiment of *Week2 Monday*. The entire experimental results are shown in Table 1.

Result analysis: As shown in Fig. 5, it is easily visible that the performance of our scheme totally outperforms the performance of Snort. At the very beginning, the completion time is hard to distinguish (i.e., when the number of rules is 50) while our scheme gradually beats the Snort in the aspect of reducing the completion time with the increase of Snort rules. For example, our scheme reduces the completion time by 25.8% when the number of rules is up to 2000. For the two-week experiment, we present the average percentage of the completion-time reduction (or called *saving time*) in Table 1. The average percentage scope of the saving time is from 16.2% to 32.5%. This statistical data reflects that our scheme is feasible and effective in improving the signature matching.

[4] The duration time needed for completing the signature matching.

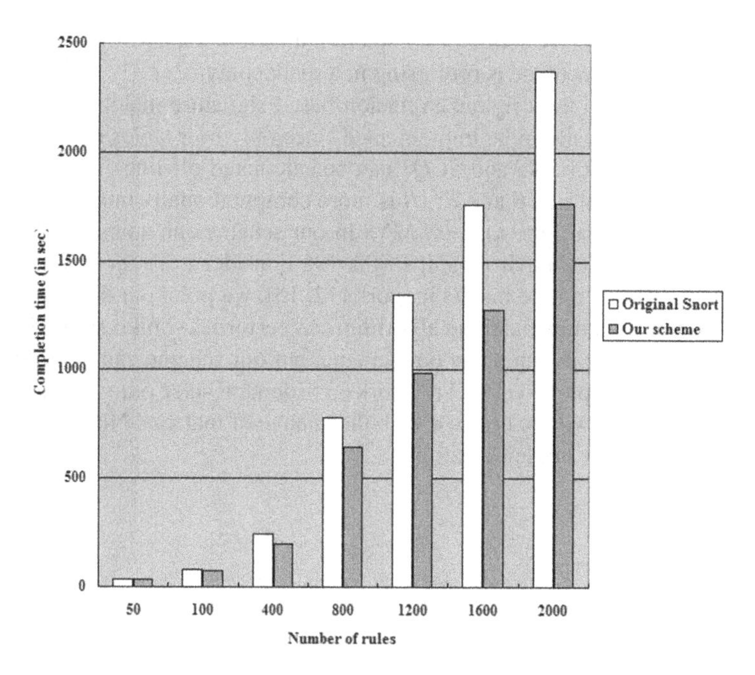

Fig. 5 The example performance of Snort and our scheme in the experiment of *Week2 Monday*.

Table 1 The average saving time by using our scheme vs. Snort in the experiment for two weeks.

Saving time (%)		
Week Day	*Week1*	*Week2*
Monday	20.0	32.5
Tuesday	27.7	21.6
Wednesday	16.2	20.8
Thursday	27.9	29.7
Friday	30.4	28.1
Saturday	28.5	30.2
Sunday	22.8	31.7

4.3 Discussion

In the experiment, we showed the initial performance of our scheme and presented the positive results. Based on the above results, we find that our scheme performs better than the original Snort since our scheme is designed based on the fact that most packet payloads will not match any NIDS signature in most cases. Thus, our scheme can quickly find a mis-match by using exclusive signature matching in achieving less completion time. According to the results in Table 1, our scheme can achieve an encouraging reduction of completion time under a network environment which shows that our scheme is promising in a real deployment.

When compared to the original exclusion-based signature matching algorithm, our scheme is more easily to be implemented since only four tables need to be pre-processed (the tables of *SNS* and *SCQ1* can be calculated off-line). The use of *occurrence bitmap* in both *ExB* and E^2xB is more computationally intensive than our approach. What is more, the *table of MNS* in our scheme can adapt itself according to different network environment so that we consider it as another advantage of our scheme. Based on the results in work [12, 13], we point out that the original exclusion-based signature matching algorithm can perform a similar statistical time-reduction in signature matching as our scheme, but our scheme can achieve better performance when deployed in real network environment since our scheme is adaptive (i.e., *SCQ2* table will be updated with the change of matched NIDS signatures) and is more simple for implementation.

5 Future Work

The initial performance of our *single character frequency-based exclusive signature matching scheme* is encouraging. But there are still some issues that we can explore and improve in future development. First, we can measure our scheme with more parameters such as bit-string length, packet size and false match rate. To explore the particular effect of these parameters on our scheme can lead us to better understand the overall performance of our scheme. Second, we plan to directly compare our scheme with the original exclusion-based signature matching to obtain more intuitive results. At last, we plan on using more datasets to evaluate our scheme and implement our scheme in a high volume traffic environment to explore and analyze its performance with the purpose of leading to further optimizations.

6 Conclusion

The poor performance of NIDSs under a high volume traffic environment is a big problem to encumber its further development. The reason is that a NIDS will spend most of its total processing time in comparing the input strings with its signatures. To mitigate this issue, a more efficient signature matching algorithm is desirable with the increasing number and complexity of NIDS signatures.

In this paper, we developed a scheme of *single character frequency-based exclusive signature matching* which is based on the observation that most packet payloads will not match any NIDS signature in most cases, aiming to improve the performance of NIDSs under heavy traffic environment by reducing the consuming time in signature matching. In particular, our scheme contains four tables and a decision component in selecting the most appropriate single character for exclusive signature matching. When a packet arrives, our scheme will choose the most appropriate single character according to a decision algorithm. In the experiment, we implemented our scheme in a network environment. The experimental results encouragingly show that our scheme outperforms the performance of original Snort by greatly reducing the completion time in signature matching.

References

1. Paxson, V.: Bro: A System for Detecting Network Intruders in Real-Time. Computer Networks 31(23-24), 2435–2463 (1999)
2. Scarfone, K., Mell, P.: Guide to Intrusion Detection and Prevention Systems (IDPS). NIST Special Publication 800-94 (February 2007)
3. Roesch, M.: Snort: Lightweight intrusion detection for networks. In: Proceedings of Usenix Lisa Conference, pp. 229–238 (1999)
4. Dreger, H., Feldmann, A., Paxson, V., Sommer, R.: Operational experiences with high-volume network intrusion detection. In: Proceedings of ACM Conference on Computer and Communications Security, pp. 2–11 (2004)
5. Vigna, G., Kemmerer, R.A.: NetSTAT: a network-based Intrusion Detection Approach. In: Proceedings of Annual Computer Security Applications Conference, pp. 25–34 (1998)
6. Sommer, R., Paxson, V.: Outside the closed world: On using machine learning for network intrusion detection. In: IEEE Symposium on Security and Privacy, pp. 305–316 (2010)
7. Valdes, A., Anderson, D.: Statistical Methods for Computer Usage Anomaly Detection Using NIDES. Technical report, SRI International (January 1995)
8. Ghosh, A.K., Wanken, J., Charron, F.: Detecting Anomalous and Unknown Intrusions Against Programs. In: Proceedings of Annual Computer Security Applications Conference, pp. 259–267 (1998)
9. Snort, The Open Source Network Intrusion Detection System, `http://www.snort.org/` (cited January 2012)
10. Rivest, R.L.: On the worst-case behavior of string-searching algorithms. SIAM Journal on Computing 6, 669–674 (1977)
11. Fisk, M., Varghese, G.: An analysis of fast string matching applied to content-based forwarding and intrusion detection. Technical Report CS2001-0670, University of California, San Diego (2002)
12. Markatos, E.P., Antonatos, S., Polychronakis, M., Anagnostakis, K.G.: Exclusion-based Signature Matching for Intrusion Detection. In: Proceedings of the IASTED International Conference on Communications and Computer Networks, pp. 146–152 (2002)
13. Markatos, K.G., Antonatos, S., Markatos, E.P., Polychronakis, M.: E2xB: A Domain-Specific String Matching Algorithm for Intrusion Detection. In: Proceedings of IFIP International Information Security Conference, pp. 217–228 (2003)
14. Boyer, R.S., Moore, J.S.: A fast string searching algorithm. Communications of the ACM 20(10), 762–772 (1977)
15. Horspool, R.: Practical fast searching in strings. Software Practice and Experience 10, 501–506 (1980)
16. Baboescu, F., Varghese, G.: Scalable packet classification. In: Proceedings of Conference on Applications, Technologies, Architectures, and Protocols for Computer Communications (SIGCOMM), pp. 199–210 (2001)
17. Aho, A.V., Corasick, M.J.: Efficient string matching: An aid to bibliographic search. Communications of the ACM 18(6), 333–340 (1975)
18. Wu, S., Manber, U.: A Fast Algorithm for Multi-Pattern Seaching. Technical Report TR-94-17, Department of Computer Science. University of Arizona (May 1994)

19. Commentz-Walter, B.: String Matching Algorithm Fast on the Average. In: Maurer, H.A. (ed.) ICALP 1979. LNCS, vol. 71, pp. 118–132. Springer, Heidelberg (1979)
20. Kim, K., Kim, Y.: A fast multiple string pattern matching algorithm. In: Proceedings of AoM/IAoM Conference on Computer Science (August 1999)
21. Lakshman, T.V., Stidialis, D.: High speed policy-based packet forwarding using efficient multi-dimensional range matching. In: Proceedings of Conference on Applications, Technologies, Architectures, and Protocols for Computer Communications, SIGCOMM, pp. 203–214 (1998)
22. Song, H., Lockwood, J.W.: Efficient packet classification for network intrusion detection using fpga. In: Proceedings of ACM/SIGDA International Symposium on Field Programmable Gate Arrays, FPGA, pp. 238–245 (2005)
23. Srinivasan, V., Varghese, G., Suri, S., Waldvogel, M.: Fast scalable level four switching. In: Proceedings of Conference on Applications, Technologies, Architectures, and Protocols for Computer Communications, SIGCOMM, pp. 191–202 (1998)
24. Gupta, P., McKeown, N.: Packet classification on multiple fields. In: Proceedings of Conference on Applications, Technologies, Architectures, and Protocols for Computer Communications, SIGCOMM, pp. 147–160 (1999)
25. Meng, Y., Kwok, L.-F.: Adaptive Context-aware Packet Filter Scheme using Statistic-based Blacklist Generation in Network Intrusion Detection. In: Proceedings of International Conference on Information Assurance and Security, IAS, pp. 74–79 (2011)
26. Sourdis, I., Dimopoulos, V., Pnevmatikatos, D., Vassiliadis, S.: Packet pre-filtering for network intrusion detection. In: Proceedings of ACM/IEEE Symposium on Architectures for Networking and Communications Systems, ANCS, pp. 183–192 (2006)

Data Prefetching for Scientific Workflow Based on Hadoop

Gaozhao Chen, Shaochun Wu, Rongrong Gu, Yongquan Xu, Lingyu Xu, Yunwen Ge, and Cuicui Song

Abstract. Data-intensive scientific workflow based on Hadoop needs huge data transfer and storage. Aiming at this problem, on the environment of an executing computer cluster which has limited computing resources, this paper adopts the way of data prefetching to hide the overhead caused by data search and transfer and reduce the delays of data access. Prefetching algorithm for data-intensive scientific workflow based on the consideration of available computing resources is proposed. Experimental results indicate that the algorithm consumes less response time and raises the efficiency.

Keywords: Hadoop, data-intensive, scientific workflow, prefetching.

1 Introduction

In recent years, large-scale science and engineering calculation are first broken into multiple complex applications which are coded by many different fields and organizations. Those applications are organized into scientific workflow by particular logistic concern [1]. Along with the rapid development of computer and network technology, scientific data and knowledge such as geography, biology and environmentology are growing by exponential order, which has a further requirement raised on analytical and processing power of scientific workflow. How to improve the service quality of data-intensive scientific workflow has become an important issue.

Gaozhao Chen · Shaochun Wu · Rongrong Gu · Yongquan Xu ·
Lingyu Xu · Yunwen Ge · Cuicui Song
School of Computer Engineering and Science, Shanghai University,
Shanghai 200072, P.R. China
e-mail: chengaozhao2008@163.com, scwu@shu.edu.cn,
gurong_rong@126.com, xuyongquan54321@126.com,
xly@shu.edu.cn, geyunwen@shu.edu.cn, scui1126@163.com

R. Lee (Ed.): Computer and Information Science 2012, SCI 429, pp. 81–92.
springerlink.com

Hadoop provides a good solution to the management and usage of massive data, its Mapreduce programming model is suitable for distributed parallel process large scale datasets [2]. The Hadoop Distributed File System (HDFS) is appropriate for operating in general hardware platform and has high fault-tolerant, it is a very effective control for deployment on cheap commodity hardware. HDFS can provide high-throughput data access. It is particularly suited for those applications which have large scale datasets [3]. Due to the effectiveness and high reliability of Hadoop, as well as the support for massive data processing, our task group designed a marine scientific workflow prototype system based on Hadoop platform through the full use of the marine information processing model and related resources. Through the integration of all aspects of resources, completely innovative, our task group's goal is to ensure the marine researchers are able to customize personalized workflow, which can effectively analyze and process massive marine data.

Based on the analysis and study of the marine data processing scientific workflow prototype system in Hadoop platform, with the background of mass data analysis and processing, a data prefetching for scientific workflow is given in this paper. It can be used to improve the data locality and I/O workloads of scientific workflow task on the environment of multi-user shared Hadoop cluster. The aim is to minimize the runtime of data-intensive scientific workflow task, thus further improving the service quality of the scientific workflow system.

2 Related Works

Scientific workflow derives from business workflow. Business workflow focuses on process and towards the control stream, while scientific workflow focuses on data and towards the data stream. A remarkable characteristic of scientific computing is data integration. It requires frequent data manipulations during the calculation, scientific computing calls specific application to analyze, migrate, or store this data on distributed machines according to certain regulations and logical relations. Its dependence focuses on the data stream between the data creator and data consumer among the subtasks of the workflow task [4]. Meanwhile, scientific workflow has the characteristic of large-scale integration. Now, the data we collected reach TB level even PB level every day, especially in the fields of geography, biology, environics and etc. A slew of professional programs need to be combined to analyze and process these mass data. Figure 1 shows an example of scientific workflow DAG.

In order to improve the service quality of workflow system, some scientific workflow system such as Pegasus, DAGMan and Taverna adopted performance-driven scheduling to minimize the runtime of workflow task [5]. Scheduling algorithm for scientific workflow based on the consideration of available storage [6]. used data jobs to forecast the available storage of the compute node, then take the optimized tasks scheduling to minimum the runtime of data-intensive scientific workflow when the cluster have limited storage resource. SWARM [7] forecasted and filled missing data of target datasets by applying Bayesian probability inference, automatically improved the data quality of metabolic networks.

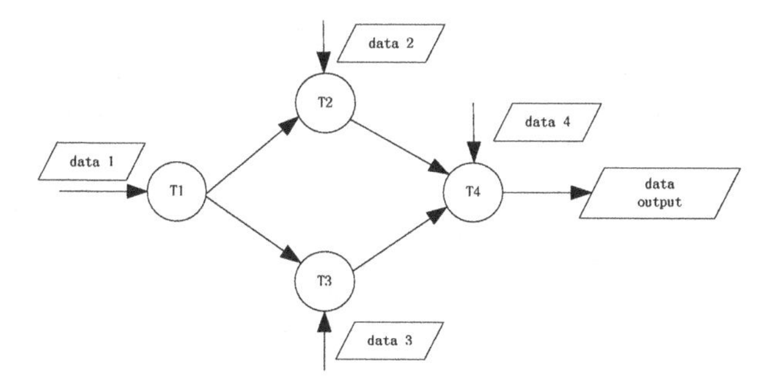

Fig. 1 An example of data-intensive scientific workflow DAG.

This paper proposes a prefetching algorithm to improve the service quality of scientific workflow system. Prefetching is one of the key technologies to hide the overhead caused by data search and transfer and reduces the delays of data access. It is widely applied in all levels of traditional computer system. Prefetching is an active service strategy. The key thought is to fetch the data objects which may be accessed in the near future by making full use of the idle periods of the system, now there are already many prefetching strategies to improve the system performance and user experience, such as Web prefetching[8], storage prefetching[9], disk array cache prefetching[10]. Most of their works focused on just file prefetching on local file systems, the applied research on distributed computing environment is still limited. This paper proposes a prefetching mechanism which applies to scientific workflow system based on Hadoop platform. In the execution of a scientific workflow task, the prefetching algorithm can prefetch the datasets which may be used in the near future, improve data locality of the applications inside the workflow task, reduce the time that the application spends waiting for data to return from the server and minimize the runtime of mass data analysis and processing scientific workflow task.

3 The Architecture of the Workflow Service Prototype System Based on Hadoop Platform

Figure 2 shows the architecture of the workflow service prototype system based on Hadoop platform. Here, the subtask and application of a workflow task is defined as model. The study of customizable composite model workflow has three components: visual modeling technique of customizable composite model service, workflow analysis and recombination, and workflow implementation based on distributed collaborative computing service. Through customizable composite model service, users can access all kinds of model service that the resource layer provides through the web service.

Users customize related service based on their own needs by means of the visual modeling tool. The model service has good adaptivity which has been fully reflected in the customizing module. The users customize personalized model service in the interactive visual interface, after normalization, the service will be delivered to the backdoor management system in the form of workflow. The backdoor management system selects an appropriate model based on the customer's service request, and then creates an effective and reasonable workflow order form by the topological relation between each model. The visual modeling tool module offers interface for upper level service and lower workflow through the visual modeling language, builds the visual modeling platform, shows the model to the users in the form of document, graphic and pictures, describes the user order form in the form of visual modeling language described code, and then delivers the code to the Hadoop platform.

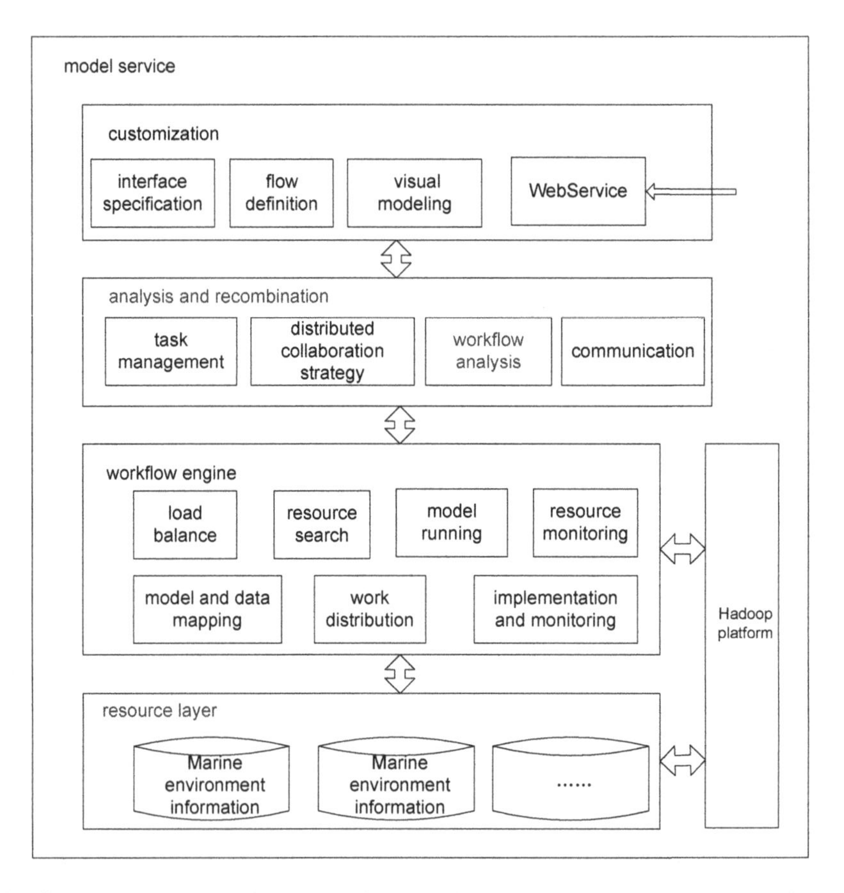

Fig. 2 The architecture of the workflow service prototype system based on Hadoop platform.

The workflow analysis and recombination module are in charge of code analysis, workflow organization and split, workflow checking and scheduling strategy. Workflow analysis module analyze each model and the relation between each model based on the code from front end, restore the customized workflow, restructure the analysis result, analyze the predecessor and successor relation according to distributed collaborative computing, and form the execution order of workflow node. The strategy of node recombination determines the submit approach of workflow node.

Workflow engine is in charge of drives and manages the workflow, and schedules the workflow task to the hadoop platform, coordinates serial and parallel execution of every model of a workflow task, checks the matching of model and data or whether the parameter between models is proper, and ensure the efficient and accurate execution of every model of a workflow task.

The marine environment data and model, as well as the user model are stored in resource layer. Most of the data of the resource layer are stored in HDFS. The resource layer can search and locate data and model by calling related functional modules.

4 Data Prefetching for Scientific Workflow Based on Hadoop

Our prefetching strategy focuses on data locality of a computing, when a workflow task has been submitted to hadoop platform by workflow engine, some computation tasks are located away from the data they consume, we think those computations have bad data locality, the data needed by computation task can be pushed to the node of the computation or other location near the node to improve data locality. The aim is to minimize the runtime of mass data analysis and processing scientific workflow task.

4.1 Description of the Problem

The traditional workflow system commonly carries out within single–site organization, the real time information of resource in total organization can be obtained centrally, and the workflow task is easy to be carried out. But the dynamic, heterogeneity and dispersivity of resource on hadoop distributed computing environment must be considered. We designed the workflow system to analyze and process mass marine data according to the project. Each application or subtask in the workflow task is used for computing corresponding large data. There is a data steam between the data creator and data consumer. When a computation task is located near the data it consumes, the task has very high efficiency in the implementation [11]. It is hardly plausible that allocates each user enough physical resource. In fact, not only the resources are shared by multi users, but also the resources that a user can use are limited. It is difficult to ensure all the computation tasks have good data locality under such circumstances. When the cluster has a bad processing performance, the execution of whole scientific workflow task will be badly affected.

The method given in this paper is used to improve data locality of computation task by prefetch the expected block replica to local rack, the prefetching runs in block level. The idea of the prefetching algorithm is shown in figure 3: when a workflow task has been submitted to the cluster, the cluster allocates a number of computing resource within rack A for the task, the workflow task is broken down into a list of jobs that are composed of multiple subtasks. When a job in the list is going to run, the data the job need are D1, D2, D3, D4, D5 and D6, D1 and D3 are stored in Node 1, D2 and D4 are stored in Node 2 and D5, D6 are stored in node m of rack B. When D1, D2, D3 and D4 in local rack A have been processed, D5, D6 in rack B or their replicas must be acquired. There are additional overhead caused by data search and transfer when request data D5, D6. If the network conditions between the two racks are not ideal, the job will experience high delay that waiting for data to return from the server, and reduce the efficiency of process execution, consequently slow down the execution speed of the whole workflow task. Obviously, the data needed by the job can be prefetch before the execution of job, and fetch D5, D6 from rack B to local rack A ahead of schedule, make every subtask of the workflow task have good data locality, thus greatly increased performance.

In order to meet the requirements of resource allocation on the environment of multi-user cluster, due to the high doncurrency between each subtask of a workflow task, In this paper, Hadoop fair scheduler is used to organize resources and schedule jobs, ensuring that the concurrent tasks obtain equal sharing resources.

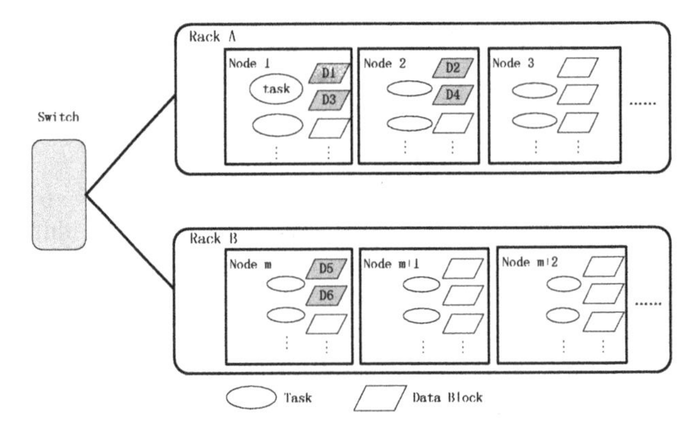

Fig. 3 The idea of prefetching algorithm that improve data locality.

4.2 *Dynamic Replica Selection*

The default method of replica selection is trying to selects closest replica for each block, this strategy causes "hot" DataNodes which frequently serve requests, causing high workload of the DataNodes. A novel dynamic replica selection algorithm is introduced in this paper. Table 1 lists the summary of notations used in the algorithm.

Table 1 Notations used in the dynamic replica selection algorithm

Notation	Description
CopyDNSet	The set of DataNodes where all replicas of a block locate
sc	The size of the data block , default is 64M
d	A DataNode in the set *CopyDNSet*
PA_d	Processing power of the node, such as CPU capability, memory capability, etc.
WL_d	Workload of the node, including CPU availability, memory availability, etc
VL_d	Read and write speed of disc in node d
BW_d	The available network bandwidth between the location of the data and data consumer

The cost of obtain a replica from DataNode includes data block replica processing and transmission cost. *COP* represents the transmission cost per a unit of data. *COT* represents the processing cost per a unit of data, using a formula expressed as:

$$
\begin{aligned}
COST_d &= COST_d^P + COST_d^T \\
&= COP \bullet \frac{\Delta S}{PA_d - WL_d} \bullet sc + COT \bullet [t + t_0] \\
&= COP \bullet \frac{\Delta S}{PA_d - WL_d} \bullet sc + COT \bullet \left[sc \bullet (\frac{1}{BW_d} + \frac{1}{VL_d}) + t_0 \right] \quad d \in CopyDNSet
\end{aligned} \tag{1}
$$

sc is the size of a data block replica of HDFS, t is data transfer time, including the time for reading data from disc and data transfer in network, ΔS is a constant, t_0 is start-up time, considering the load balance of HDFS, we must choose the most suitable node and get the needed data replica from the node to minimize the total cost.

$$SCopy = \min\{COST_d\}, d \in CopyDNSet \tag{2}$$

When DataNodes join in the HDFS cluster, they will register their information with NameNode, including the processing power of the node, such as CPU capability, memory capability, etc. The available network bandwidth between the location of the data and data consumer, as well as the workload of a node must be obtained in real time.

4.3 Prefetching Algorithm

There is a need for data transmission when tasks are located away from the data they consume, which make the job experience high delay, it could even lead to failure. Those tasks need run again under the control of JobTracker, result in high runtime of the job and then slow the execution of the whole workflow task.

A workflow task is parsed to many serial jobs when it has been submitted to hadoop platform, a job consists of one or more subtasks, and multiple tasks in a job are highly parallelizable. The prefetching algorithm is running in the workflow engine, and prefetch the expected block replica which needed to the subtask to local rack, makes the data needed to the task near the TaskTracker, and reduces the time spent in the application waiting for data transmission when task run on TaskTracker. The key thought is to fetch the data objects which may be accessed in the near future by taking full advantage of the idle periods of system, and improves data locality of computation task. The algorithms are described in figure 4 and figure 5.

```
Input: a workflow task wt
QuickRun(WorkFlowTask wt)
{
   Parse the users' workflow task wt, put all the subtasks into
   a job list JL, a subtask may be consisted of several tasks;
   Set the initial prefetch range m;
   For each job j in job list JL
      Run job j, in the mean time, activates
      the prefetch process, prefetch data for the following
      m jobs by using CrossRackPrefetch algorithm;
      If ( the job j has been completed)
         If (the current system workload is above the setting
           threshold)
           decreases the prefetch range m;
            else
           increases the prefetch range m;
            End if
         End if
   End for
}
```

Fig. 4 The execution of a workflow task with prefetching

In HDFS, the block information which constitutes a file can be obtained from Namenode by using Remote Procedure Calls (RPC). For each block, Namenode return the addresses of DataNodes which contains the block. The addresses of DataNodes which contains the block's replica can be obtained by the triplets attribute of BlockInfo. The location information of all blocks and replicas of a data file is used to guide prefetch.

The prefetching could not wholly perform its functions for being too conservative or too positive, real time workload of the cluster need to be comprehensively considered, we set an initial prefetching range, get the real time working condition of computing node from monitoring module of the workflow system. When real time workload is below the setting threshold, it increases the prefetch range; otherwise, it decreases the prefetch range. It ensures that the prefetching range is adequate by adjusting it dynamically and does no harm to the overall performance of the cluster.

```
Input: a list of jobs jlist
CrossRackPrefetch(JOBLIST jlist)
{
   For each job j in jlist
      Get the all of block information which constitute the
       input of a job, the information  is stored in DBL;
      For each data block d in DBL
        If (d and its replicas are not in local rack R)
           Choose a node N with enough space in the rack
        R;
           If (N. AvailableSpace< N.Length)
              break;
           End if
           According to formula(1) (2) mentioned above,
        choose the optimal node NB_min which contains d's
        replica;
           Prefetch the data block replica from NB_min to
        N;
           N. AvailableSpace = N. AvailableSpace –
        N.Length;
           Update the list DBL;
        End if
```

Fig. 5 Description of the prefetching algorithm

5 Experiments and Analysis

We design the experiments according to the proposed improvement. The test platform is built on a cluster of 8 PC servers and 3 switches. In each node, Ubuntu server 10.10 with the kernel of version 2.6.35-server is installed. Java version is 1.6.27 and Hadoop version is 0.20.203. The size of HDFS blocks is 64 MB and the number of replicas is set to 3.

In the experimental system, the process that user submit workflow task is shown in figure 6, the user customize the scientific workflow task through browser, the task is submitted to the web server, then through the request controller, the task is delivered to the workflow system and to be resolved. Finally, the task is distributed to the hadoop cluster, and a number of computing node is assigned to process the task.

In order to analyze the effect of the algorithm combine the dynamic replica selection and prefetching, this paper takes three different approaches to schedule and execute a data-intensive scientific workflow task.

1) Algorithm A1: the algorithm which combined the default dynamic replica selection and without prefetching
2) Algorithm A2: the algorithm which combined the default dynamic replica selection and prefetching
3) Algorithm A3: the algorithm which combined the dynamic replica selection and prefetching

A data-intensive scientific workflow task which contains 20 subtasks is used in these simulations, the task need to process two large test datasets, DS-1 and DS-2, the size of the datasets are 5G and 10G, and the prefetching range m are set to 1, 2, 3 and 4, each workflow task will be executed twice in the same configuration, then record the average execution time. The datasets are distributed homogeneously in the cluster, and are evenly distributed to each subtask of the workflow task as input. Given the experimental comparison, the numbers of available computing nodes *NA* are set to 2 and 4. The runtime information of the scientific workflow task in different algorithm is shown in figure 7.

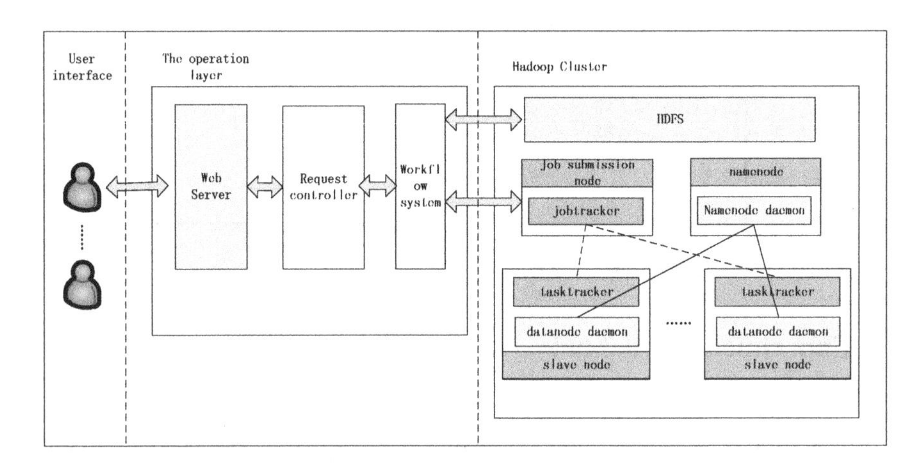

Fig. 6 The process that user submit workflow task

It can be seen from the figure 7 that the prefetching improves the efficiency of workflow task, and the larger the test dataset, the greater the effect. Meanwhile, the prefetch range settings also influence the runtime of task. When the system has enough available resources, the prefetch range is higher, and the runtime of the workflow task is shorter. Excessive prefetching will affect the ultimate performance of the system when the workload of system is too high. The dynamic replica selection algorithm also improves the efficiency of workflow task upon the basic of prefetching.

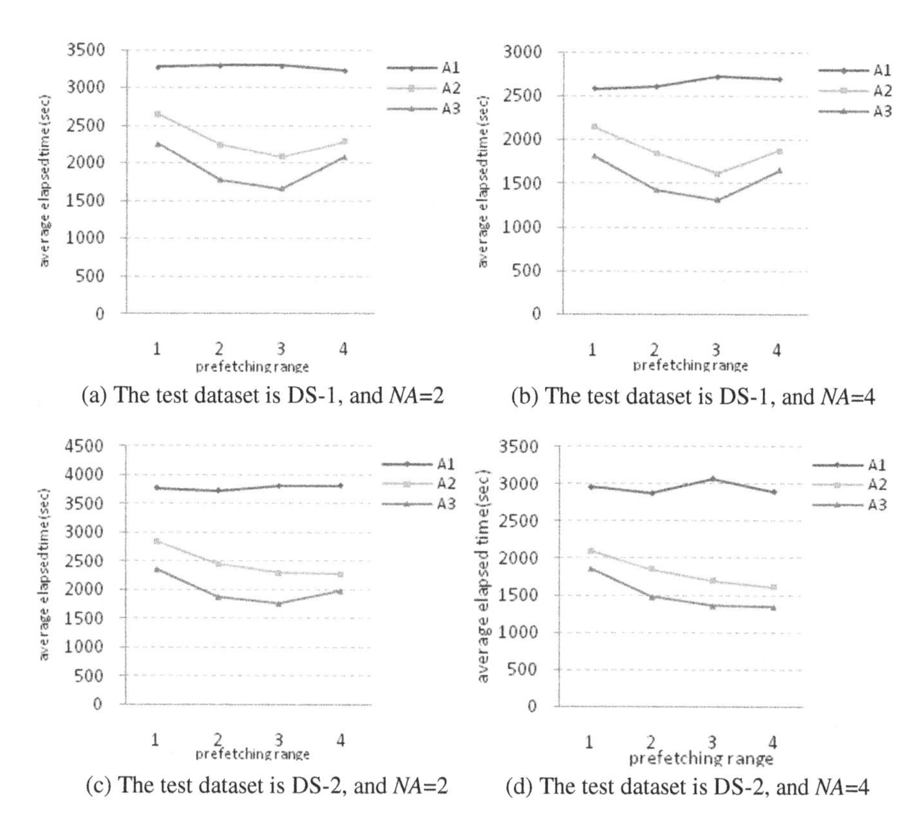

(a) The test dataset is DS-1, and *NA*=2

(b) The test dataset is DS-1, and *NA*=4

(c) The test dataset is DS-2, and *NA*=2

(d) The test dataset is DS-2, and *NA*=4

Fig. 7 The runtime information of the scientific workflow task in different algorithm.

6 Conclusions

In this paper, the prefetching with a combination of dynamic replica selection was introduced to the scientific workflow application system based on the Hadoop platform, it is a good way to improve the system efficiency and reduce the impact on efficiency due to the limited system resources. However, the efficiency of the algorithm depends on the initial setting of prefetching range. Further studies of the impact of prefetch range setting on the efficiency are needed. Furthermore, the timing of prefetching and continuity of data require further investigation.

Acknowledgment. This work was financially supported by the Ocean Public Welfare Project of The Ministry of Science and Technology under Grant No.201105033, the National Natural Science Foundation of China. (No.40976108), Shanghai Leading Academic Discipline Project (Project Number: J50103), Shanghai University Fifth Graduate students Innovation Fund (Project Number: SHUCX112164).

References

1. Taylor, I., Deelman, E., Gannon, D., et al.: Workflow in E-science. Springer, Heidelberg (2007)
2. Dean, J., Ghemawat, S.: Map/Reduce: Simplified Data Processing on Large Clusters. In: OSDI 2004:Sixth Symposium on Operating System Design and Implementation (2004)
3. `http://hadoop.apache.org/`
4. Bharathi, S.: Characterization of scientific workflows. In: Workflows in Support of Large-scale Science (s.n.), pp. 1–10 (2008)
5. Yu, J., Buyya, R.: A Taxonomy of workflow management systems for grid computing. Journal of Grid Computing 3(3-4), 171–200 (2005)
6. Tang, X., Hao, T.: Schedule algorithm for scientific workflow based on limited available storage. Computer Engineering 35, 71–73 (2009)
7. Shi, X., Stevens, R.: SWARM:A Scientific Workflow for Supporting Bayesian Approaches to Improve Metabolic Models. In: Proceedings of the 6th International Workshop on Challenges of Large Applications in Distributed Environments, pp. 25–34. ACM, New York (2008)
8. Nanopoulos, A., Katsaros, D., Manolopoulos, Y.: A data mining algorithm for generalized Web prefetching. IEEE Trans. on Knowledge and Data Engineering 15(5), 1155–1169 (2003)
9. Chen, J., Feng, D.: An intelligent prefetching strategy for a data grid prototype system. In: IEEE International Conference on Wireless Communications, Networking and Mobile Computing, Wuhan, China (2005)
10. Dominique, T.: Improving Disk Cache Hit-Ratios Through Cache Partitioning. IEEE Trans. on Computers 41(6), 665–676 (1992)
11. Seo, S., Jang, I., Woo, K., Kim, I., Kim, J.-S., Maeng, S.: Hpmr: Prefetching and pre-shuffling in shared mapreduce computation environment. In: IEEE CLUSTER (2009)

An NMF-Based Method for the Fingerprint Orientation Field Estimation

Guangqi Shao, Congying Han, Tiande Guo, and Yang Hao

Abstract. Fingerprint orientation field estimation is an important processing step in a fingerprint identification system. Orientation field shows a fingerprint's whole pattern and globally depicts the basic shape, structure and direction. Therefore, how to exactly estimate the orientation is important. Generally, the orientation images are computed by gradient-based approach, and then smoothed by other algorithms. In this paper we propose a new method, which is based on nonnegative matrix factorization (NMF) algorithm, to initialize the fingerprint orientation field instead of the gradient-based approach. Experiments on small blocks of fingerprints prove that the proposed algorithm is feasible. Experiments on fingerprint database show that the algorithm has a better performance than gradient-based approach does.

Keywords: NMF, base matrix, fingerprint orientation field, PSNR.

1 Introduction

Among many biometric recognition technologies, fingerprint recognition is most popular for personal identification due to the uniqueness, universality, collectability and invariance. A typical automatic fingerprint recognition system (AFIS) often includes the below steps: fingerprint acquisition, image preprocessing (such as fingerprint segmentation, enhancement and the orientation field estimation), fingerprint classification, minutiae detection and matching [1]. See Fig.1 for the flowchart of conventional fingerprint recognition algorithms.

As showed in the flow chart, fingerprint orientation field estimation is an important processing step in a fingerprint identification system. Orientation field shows the whole pattern of a fingerprint and globally depicts the basic shape, structure and

Guangqi Shao · Congying Han
School of Mathematical Sciences, Graduate University of Chinese Academy of Sciences Beijing, China
e-mail: guangqi-002@163.com, hancy@gucas.ac.cn

Tiande Guo · Yang Hao
Graduate University of Chinese Academy of Sciences, Beijing, China
e-mail: tdguo@gucas.ac.cn, haoyang10@mails.gucas.ac.cn

R. Lee (Ed.): Computer and Information Science 2012, SCI 429, pp. 93–104.
springerlink.com

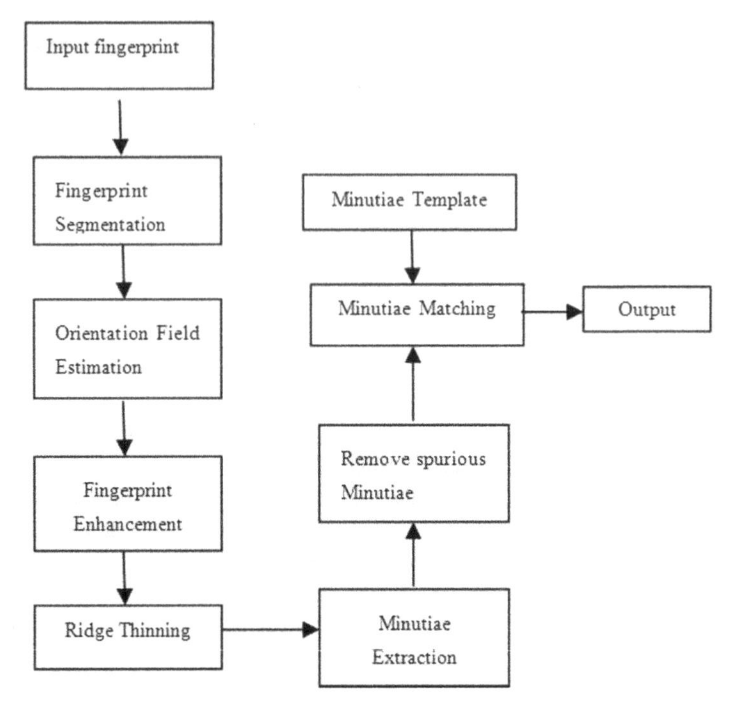

Fig. 1 Flowchart of the Fingerprint Recognition System.

direction of a fingerprint. In a fingerprint identification system, orientation field is very important and many other fingerprint processing steps are based on the orientation field, such as fingerprint enhancement [2, 3], fingerprint segmentation [4], singular point detection [5, 6], fingerprint classification [7]. Moreover, good orientation field can improve the performance of a fingerprint recognition system.

Many methods for the fingerprint orientation field estimation have been developed due to its importance. There are gray-based approaches [8, 9], gradient-based approaches [10, 11] and model-based approaches [12, 13]. The simplest and most natural approach for extracting local ridge orientation is gradient-based method. It is well known that the gradient phase angle denotes the direction of the maximum intensity change. Therefore, the direction of a hypothetical edge is orthogonal to the gradient phase angle. Based on the above idea, Rao [10], Ratha, Chen, and Jain [11] proposed an effective method for computing the fingerprint orientation field. Sherlock and Monro [12] proposed a so-called zero-pole model for orientation field based on singular points, which takes zero and the delta as a pole in the complex plane. Vizcaya and Gerhardt [13] improved the zero-pole model by using a piecewise linear approximation around singularities to adjust the zero and pole's behavior. They also proposed an optimization technique, based on gradient descent, to determine the model parameters starting from an estimation of the orientation image. In addition to the above methods, there are also filter-bank based approaches [14], methods based on high-frequency power in 3-D space [15].

In this paper, we propose a novel orientation field estimation method. In this method, we use nonnegative matrix factorization (NMF) algorithm to train matrixes from different orientations to obtain base matrixes. These base matrixes can be used to estimate fingerprint orientation field. Process as follows: First, divide the fingerprint into small blocks. Factorize each block in all base matrixes. Different representations of the block under different base matrixes are obtained. Second, recover the block using these representations under these base matrixes. Third, calculate the PSNR between the block and the recovered blocks. The orientation of the block is the orientation of base matrix corresponding to the largest PSNR. The remainder of this paper is organized as follows. Section 2 introduces the theory of NMF which includes the origin of the NMF, the method of solving it. How to use NMF to obtain the orientation field is presented in Section 3. The experimental results are exhibited in Section 4. Finally, Section 5 draws some conclusions for this paper.

2 The Theory of Nonnegative Matrix Factorization

Nonnegative Matrix Factorization was first introduced by Tapper and Paatero in [16]. But it is widely known by the works of Lee and Seung [17]. In that paper, they give two algorithms, which are effective to learn parts of the faces and semantic features. Since then, the NMF has been widely used in many fields, such as pattern recognition, signal processing and data mining.

Given an $m \times n$ nonnegative matrix $A(A_{ij} \geq 0)$ and a reduced rank $r(r \leq \min(m,n))$, the nonnegative matrix factorization problem is to find two nonnegative matrices $U(U \in R_{+}^{m \times r})$ and $V(V \in R_{+}^{n \times r})$ that approximate A as follows:

$$A \approx UV^T \tag{1}$$

There are many ways to measure the difference between the data matrix A and the matrix UV^T. But the most used measure is the Frobenius norm in the following:

$$F(A,UV^T) = \frac{1}{2}||A - UV^T||_F^2 = \frac{1}{2}\sum_{i=1}^{m}\sum_{j=1}^{n}(A_{ij} - [UV^T]_{ij})^2 \tag{2}$$

$[UV^T]_{ij}$ represents the element of matrix UV^T in row i and column j .Therefore the above problem can be written in the following style:

$$\min_{U \in R_{+}^{m \times r}, V \in R_{+}^{n \times r}} \frac{1}{2}||A - UV^T||_F^2 \tag{3}$$

Moreover,the problem can be extended to include auxiliary constraints on U and V. This can be used to enforce the desired properties in the solution. Penalty terms are typically used to enforce the constraints, as follows:

$$\min_{U \in R_{+}^{m \times r}, V \in R_{+}^{n \times r}} \frac{1}{2}||A - UV^T||_F^2 + \alpha_1 f_1(U) + \alpha_2 f_2(V) \tag{4}$$

Where $f_1(U)$ and $f_2(V)$ are the penalty terms; α_1 and α_2 are parameters that balance the trade-off between the main part and the constraints.

Generally, the objective function is non-convex on U and V . However, suppose that U is fixed, the function $\frac{1}{2}||A - UV^T||_F^2$ can be seen as a composition of the Frobenius norm and a linear transformation of V. Therefore, $\frac{1}{2}||A - UV^T||_F^2$ is convex on V now. In the similar, if V is fixed, the function is convex on U.

Through it is difficult to obtain the global optimal solution; there are many methods to get the local optimal solution or stable point. Next, two of them are showed.

2.1 Multiplicative Update Rules

The most popular algorithm for the NMF problem is the multiplicative rules proposed by Lee and Seung [17]. The multiplicative update rules run as follows:

$$U^{k+1} = U^k \circ \frac{[AV^k]}{[U^k(V^k)^T V^k)]} \tag{5}$$

$$V^{k+1} = V^k \circ \frac{[A^T U^{k+1}]}{[V^k(U^{k+1})^T U^{k+1}]} \tag{6}$$

The symbol"$\circ$" represents the Hadamard product, whose properties can be found in [18].

$\frac{[\cdot]}{[\cdot]}$represents the Hadamard division.

Lee and Seung has proven that the $\frac{1}{2}||A - UV^T||_F^2$ is non-increasing under the updating rules of the algorithm. However, the proof of the convergence was wrong in the original paper [17]. Lin [20] has proof it using a projected gradient bound-constrained optimization method.

2.2 Rank-One Residue Iteration

Rank-one residue iteration is a new algorithm proposed by Ngoc-Diep Ho in [19]. Let u_i and v_j be respectively the columns of U and V , and then the NMF problem can be written as follows:

$$\min_{u_i \geq 0, v_i \geq 0} \frac{1}{2}||A - \sum_{i=1}^{r} u_i v_i^T||_F^2 \tag{7}$$

If we fix all the variables except for the single vector v_t , the above problem will become:

$$\min_{v \geq 0} \frac{1}{2}||R_t - u_t v^T||_F^2 \tag{8}$$

Where $R_t = A - \sum_{i \neq t} u_i v_i^T$

The detail of the algorithm can be seen in literature [19].

Fig. 2 100 small blocks about 90 degree.

3 NMF-Based Method for Orientation Field Estimation

In this section, we will talk about how to use the NMF algorithm to calculate fingerprint orientation field. The method we propose includes three parts: choose the training set, train the fingerprints to get the base matrixes and calculate the orientation field.

3.1 How to Choose the Training Set

The local ridge orientation at a pixel is the angle that the fingerprint ridges, crossing through a small neighborhood centered at the pixel, form with the horizontal axis. The orientation of the fingerprint ranges from 0 to 180 degree.

There are two steps to obtain the training images. First, quantity the interval $[0,\cdots,180]$, namely, divide the interval $[0,\cdots,180]$ into K equal intervals. Each interval be represented by an orientation. Second, cut the whole sample fingerprints into small blocks. The orientation consistency of these small blocks is calculated. For every interval, choose small blocks whose orientations lie in the interval and have a higher orientation consistency. Every interval has the same number of small blocks. Fig.2 shows 100 small blocks which are arrayed into an image. In the next step, these small blocks will be trained.

3.2 How to Get Base Matrixes

Because the structure information of images is based on the content of the images, we can't deal with them like common matrixes. Special treatments are needed. Suppose the size of small blocks is $s \times t$ .For each interval, small blocks in the interval are arranged into column vectors. These column vectors are put together. Then, a matrix can be obtained as follows:

$$A = [A_1, A_2, \cdots, A_m] \tag{9}$$

Where $A_i \in R_+^{s\times t,1}$, $i = 1, \cdots, m$ is obtained by arranging the ith small block into a column and $A \in R_+^{s\times t,m}$.

The NMF algorithm is applied on the matrix A to get the base matrix L and the coefficient matrix $B = [B_1, B_2, \cdots, B_m]$, which satisfies the equation $A \approx LB, L \in R_+^{s\times t,r}, B \in R_+^{r,m}$. In fact, each column in L represents an image called base image. After the factorization, r base images are obtained. For original small block A_i , it is represented as LB_i now.

For each interval, the similar operations are employed. Because there are K intervals from section A, K base matrixes are obtained denoted by $L_i, L = 1, \cdots, K$.

3.3 How to Calculate the Orientation Field

For a fingerprint, divide it into small blocks. For each block X and base matrix $L_i, i = 1, \cdots, K$, the following operations are done.

$$Y_i = L_i^+ X \tag{10}$$

$$X' = [L_i Y_i]_+ \tag{11}$$

Where L_i^+ is generalized inverse L_i; $[\cdot]_+$ is the operator that makes negative elements become 0 while positive elements are unchanged.

The formula 10 is used to calculate the projection X on the ith base matrix L_i. The formula 11 recover X from X'. In order to measure the difference between the original small block X and the recovery matrix X', calculate the PSNR:

$$PSNR = 10 log_{10} \frac{255^2}{MSE} \tag{12}$$

$$MSE = \frac{||X - X'||_F^2}{s \times t} \tag{13}$$

Because there are K intervals, K PSNR will be obtained. Choose the orientation of the interval according to the largest PSNR as the orientation of the block. The reason that we choose the orientation of the largest PSNR is that the base matrixes of the similar images are most suitable for the block. In this way, we can obtain the orientations of all blocks. Therefore the orientation field of the input fingerprint is obtained.

4 Experiment Results

In this section, we give the experiments of the proposed method. The experiment includes two parts. The first part is used to prove the feasibility of the new algorithm. The second is conducted on fingerprint database.

4.1 The Feasibility of the New Method

In this part, the proof that our thought is right is given. We only need to proof that the orientation of the block corresponds to the orientation of the interval with the largest PSNR between the original block and the recovered block.

1200 small blocks of fingerprints are produced. The interval $[0, 180]$ is divided into four parts. There are 300 blocks for each interval. The size of every block is 32×32. In the experiments, we use the following multiplicative updates rules in the learning for base matrixes.

$$U^{k+1} = U^k \circ \frac{[AV^k]}{[U^k(V^k)^T V^k)] + \varepsilon} \tag{14}$$

$$V^{k+1} = V^k \circ \frac{[A^T U^{k+1}]}{[V^k(U^{k+1})^T U^{k+1}] + \varepsilon} \tag{15}$$

where ε is a small number, in order to avoid the case that the denominator is 0.

The convergence condition is

$$||A - UV^T||_F^2 \leq 50 \tag{16}$$

If the train is unfinished within 10000 iterations, the procedure stops. There are many literatures about how to choose the initial base matrixes. For simplicity, the initial base matrixes are set randomly. By training, four pairs of base matrixes are obtained. Next, four groups of small blocks are tested.

Fig.3 shows a portion of training images about 45 degree. There are 100 small blocks in Fig.3. These images are just a part of 300 small blocks. The base images in Fig.4 are corresponding to the group showed in Fig.3.

An example is given. There are 25 blocks coming from 0 degree, as showed in Fig.5. Fig.6 shows the recovered images. It is obvious that fig.6(a) is the best among them. It is because that Fig.6(a) is recovered from the base matrix about 0 degree. Table 1 reflects this point better. Test0, Test45, Test90 and Test135 represent the test sets corresponding to 0 degree, 45 degree, 90 degree and 135 degree, respectively. Base0, Base45, Base90 and Base135 represent the base matrixes corresponding to 0 degree, 45 degree, 90 degree and 135 degree, respectively. The numbers in the table represent the mean of the PSNR. The red numbers show that the largest PSNR corresponds to the base matrix of the correct orientation. So far, we prove that our new method is correct.

Fig. 3 100 small blocks of a training set.

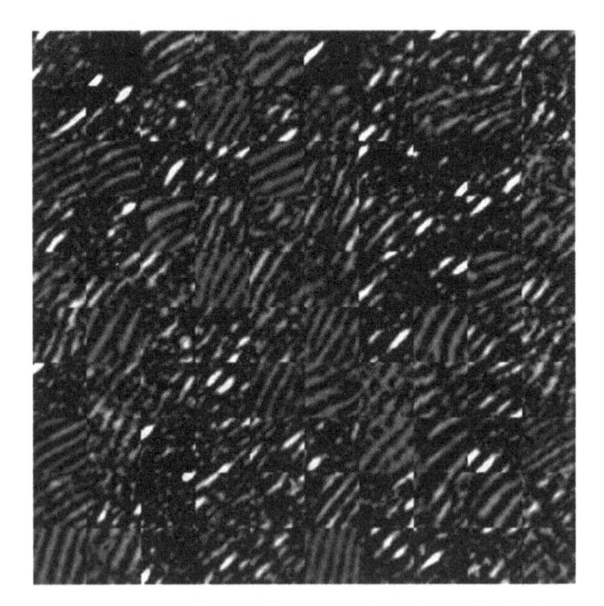

Fig. 4 The base images corresponding to the orientation showed in fig 3.

Table 1 The Results of the Experiment

PSNR Mean	Base0	Base45	Base90	Base135
Test0	31.56	24.27	21.43	24.35
Test45	25.23	30.00	24.01	23.13
Test90	21.30	24.17	31.59	24.06
Test135	23.30	20.70	21.41	28.94

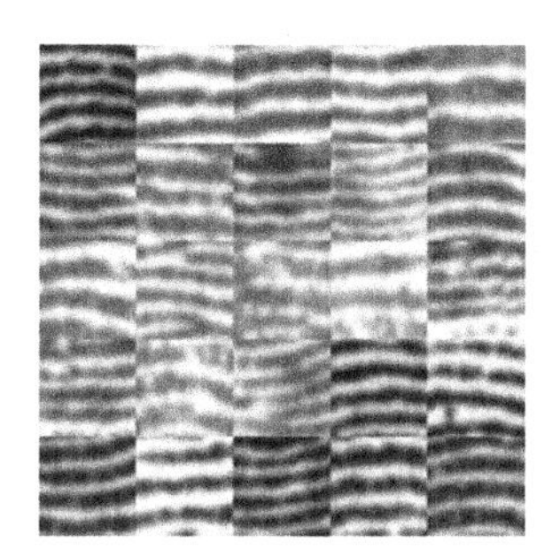

Fig. 5 25 small blocks from 0 degree

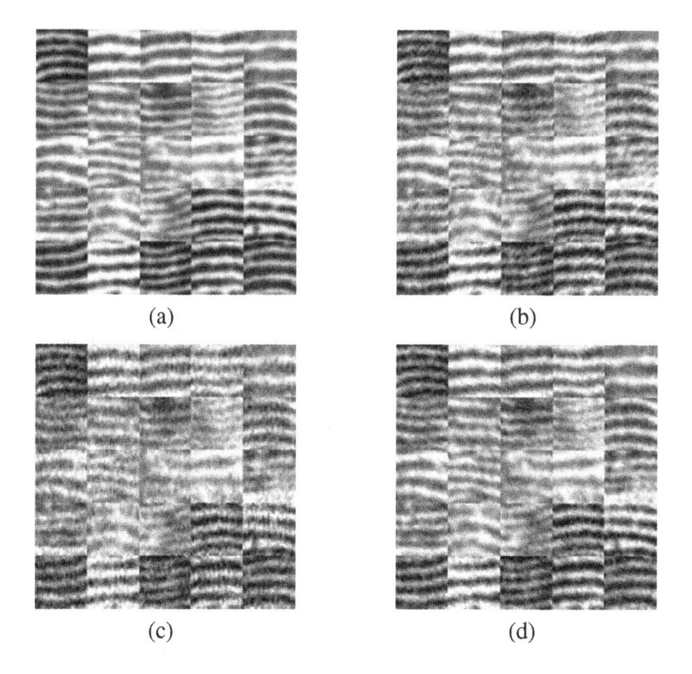

(a) (b) (c) (d)

Fig. 6 6(a), 6(b), 6(c) and 6(d) show the images showed in Fig.6 recovered from the base matrix corresponding to 0 degree, 45 degree, 90 degree and 135 degree, respectively.

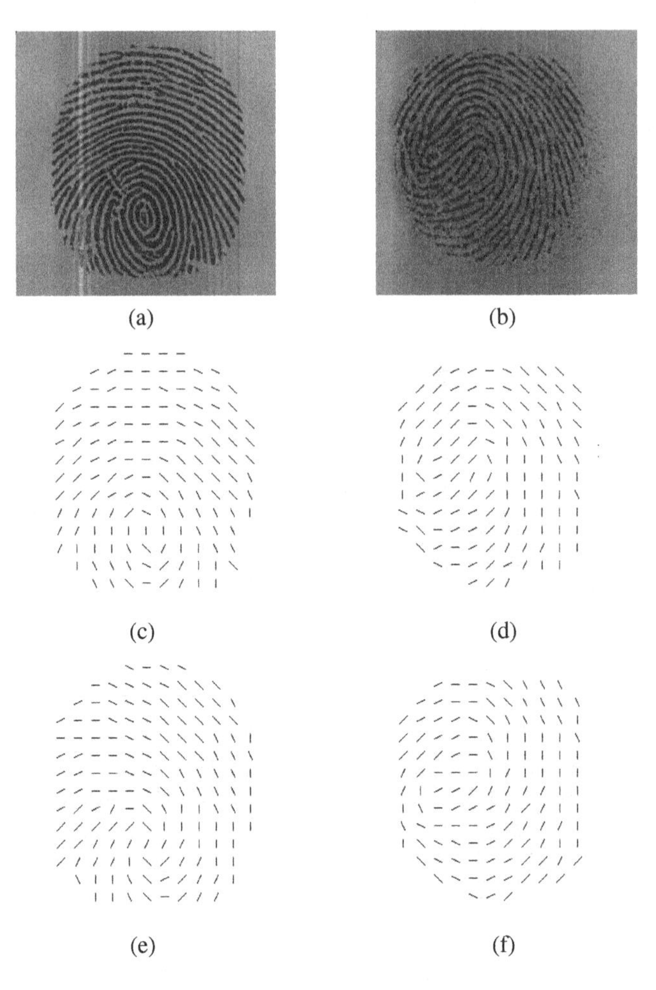

Fig. 7 a and b are original images; c and d are the orientation images estimated by our method; e and f are the orientation images estimated by the gradient-based method.

4.2 Results on the Fingerprint Database

We test the NMF-based method on FVC 2002. The size of the fingerprints is 300×300. The size of small blocks is 20×20. We set $r = 80$ and $K = 8$. The result shows that the algorithm has a good performance in practice. Moreover, we compare our method with the gradient-based method. The result reflects that our method handles the fingerprints with bad quality better. Fig.7highlights this very well. In regions with low quality, the orientation estimated by the gradient-based is wrong. However, our method can calculate it correctly.

5 Conclusion

In this paper we propose a new method for fingerprint orientation field estimation. First, we give an introduction to NMF algorithm and how to use NMF to calculate the orientation of fingerprint. In the main part, we firstly use NMF algorithm to obtain base matrixes and segment fingerprint into small blocks. Then, factorize the small blocks in all base matrixes and restore the factorized matrixes to calculate the PSNR. The orientation of the small block is the orientation of base matrix corresponding to the largest PSNR. Two experiments are tested. Our proposed method is not only feasible, but also has a good performance.

The future work will include the following aspects.

- In the training, different algorithms can be applied. Maybe there exists other kind of NMF algorithm which is most suitable for the fingerprint orientation field estimation. In the experiment, the initial matrixes are set by randomly. Though it is simple, the base matrixes may not converge. In the future experiment, the way how to choose the initial base matrixes should be studied.
- The method we propose is initiating the fingerprint orientation field. After the orientation field is obtained, some post-processing should be done as other methods do, for some bad blocks may be assigned wrongly.

Acknowledgement. This work was supported by Innovation Program of CAS, under Grant kjcx-yw-s7 and National Science Foundation of China, under Grant No.10831006 and No.11101420.

References

1. Jain, A., Hong, L., Bolle, R.: On-line fingerprint verification. IEEE Trans. Pattern Anal. Mach. Intell. 19(4), 302–313 (1997)
2. Hong, L., Wan, Y., Jain, A.: Fingerprint image enhancement: algorithm and performance evaluation. IEEE Trans. Pattern Anal. Mach. Intell. 20(8), 777–789 (1998)
3. Yun, E.K., Cho, S.B.: Adaptive fingerprint image enhancement with fingerprint image quality analysis. Image Vision Comput. 24(1), 101–110 (2006)
4. Zhu, E., Yin, J., Hu, C., Zhang, G.: A systematic method for fingerprint ridge orientation estimation and image segmentation. Pattern Recognition 39(8), 1452–1472 (2006)
5. Bazen, A.M., Gerez, S.H.: Systematic methods for the computation of the directional fields and singular points of fingerprints. IEEE Trans. Pattern Anal. Mach. Intell. 24(7), 905–919 (2002)
6. Parka, C.H., Leeb, J.J., et al.: Singular point detection by shape analysis of directional fields in fingerprints. Pattern Recognition 39(5), 839–855 (2006)
7. Wang, S., Zhang, W.W., Wang, Y.S.: Fingerprint classification by directional fields. In: Proceedings of Fourth IEEE International Conference on Multimodal Interfaces, pp. 395–399 (October 2002)
8. Kawagoe, M., Tojo, A.: Fingerprint Pattern Classification. Pattern Recognition 17(3), 295–303 (1984)
9. Mehtre, B.M., Murthy, N.N., Kapoor, S.: Segmentation of Fingerprint Images Using the Directional Image. Pattern Recognition 20(4), 429–435 (1987)

10. Rao, A.R.: A Taxonomy forTexture Description and Identification. Springer, New York (1990)
11. Ratha, N.K., Chen, S.Y., Jain, A.K.: Adaptive flow orientation-based feature extraction in fingerprint images. Pattern Recognition 28(11), 1657–1672 (1995)
12. Sherlock, B.G., Monro, D.M.: A model for interpreting fingerprint topology. Pattern Recognition 26(7), 1047–1055 (1993)
13. Vizcaya, P.R., Gerhardt, L.A.: A nonlinear orientation model for global description of fingerprints. Pattern Recognition 29(7), 1221–1231 (1996)
14. Jain, A., Prabhakar, S., Hong, L.: A multichannel approach to fingerprint classification. IEEE Trans. Pattern Anal. Machine Intell. 21(4), 348–359 (1999)
15. O'gorman, L., Nickerson, J.V.: An approach to fingerprint filter design. Pattern Recognit. 22(1), 29–38 (1989)
16. Paatero, P., Tapper, U.: Positive matrix factorization: a nonnegative factor model with optimal utilization of error estimates of data values. Environmetrics 5(1), 111–126 (1994)
17. Lee, D.D., Seung, H.S.: Learning the parts of objects by nonnegative matrix factorization. Nature (1999)
18. Horn, R.A., Johnson, C.R.: Topics in matrix analysis. Cambridge University Press (1991)
19. Ho, N.-D.: Nonnegative Matrix Factorization Algorithms and Applications (June 2008)
20. Lin, C.J.: On the convergence of multiplicative update algorithms for non-negative matrix factorization. Technical Report Information and Support Services Technical Report. Department of Computer Science, National Taiwan University (2005)

Symbolic Production Grammars in LSCs Testing

Hai-Feng Guo* and Mahadevan Subramaniam

Abstract. We present LCT_{SG}, an LSC (Live Sequence Chart) consistency testing system, which takes LSCs and symbolic grammars as inputs and performs an automated LSC simulation for consistency testing. A symbolic context-free grammar is used to systematically enumerate continuous inputs for LSCs, where symbolic terminals and domains are introduced to hide the complexity of different inputs which have common syntactic structures as well as similar expected system behaviors. Our symbolic grammars allow a symbolic terminal to be passed as a parameter of a production rule, thus extending context-free grammars with context-sensitivity on symbolic terminals. Constraints on symbolic terminals may be collected and processed dynamically along the simulation to properly decompose their symbolic domains for branched testing. The LCT_{SG} system further provides either a state transition graph or a failure trace to justify the consistency testing results. The justification result may be used to evolve the symbolic grammar for refined test generation.

1 Introduction

Formal modeling and specification languages, such as Sequence Charts in UML [20, 19], Message Sequence Charts (MSCs) [12] and Live Sequence Charts (LSCs) [4, 11], have been explored significantly to support the development of communication and reactive systems. Effective and efficient methodologies using these modeling languages in a formal testing environment [1, 14, 22] are urgently needed to support the large-scale system development. We present a formal testing tool for LSCs because LSCs have been introduced as a visual scenario-based modeling and specification language for reactive systems, with more expressive and semantically rich features compared to other alternatives [4, 8].

Hai-Feng Guo · Mahadevan Subramaniam
Department of Computer Science, University of Nebraska at Omaha,
Omaha, NE 68182, USA
e-mail: haifengguo@unomaha.edu, msubramaniam@unomaha.edu

* Corresponding author.

R. Lee (Ed.): Computer and Information Science 2012, SCI 429, pp. 105–120.
springerlink.com

Consistency checking [9, 22, 14] is one of the major and formidable problems on LSCs. In a complicated system consisting of many LSCs, inconsistency may be raised by inherent contradiction among multiple charts or due to inappropriate environmental/external event sequences. The consistency of LSCs has been shown in [9] to be a necessary and sufficient condition for the existence of its corresponding object system. Ensuring consistency of LSCs is traditionally reduced to a problem whether a satisfying object system, in a form of automaton, can be synthesized from the LSCs. Most of the consistency checking work [22, 2, 15] has been aimed at performing static analysis of LSCs using formal verification tools, and may suffer the complexities caused by transformation itself, automata synthesis, or the exponentially blowing size of transformed results [24, 10].

In this paper, we present an LSC consistency testing system, named LCT_{SG}[1], which takes LSCs and symbolic grammars as inputs and performs an automated scenario simulation for consistency testing. A symbolic context-free production grammar is used to systematically enumerate continuous inputs for LSCs, and symbolic terminals and domains are introduced in grammars to hide the complexity of different inputs which have common syntactic structures as well as similar expected running behaviors. Test generation using symbolic grammars has been shown effective for automatic test generation [16]. It is a compromise between *enumerative* and *symbolic* strategies. The enumerative test generation [17, 21, 7] may be too large and time consuming to complete, and often contain a huge number of redundant tests with same running behaviors. The *symbolic* [3, 13, 25] test generation collects symbolic constraints along the execution path, and uses a constraint solver to generate test inputs that satisfy the constraints; however, the constraint solving techniques are often expensive and domain-specific. The test generation strategy using symbolic grammar restricts a symbolic notation to apply only on terminals, while the syntactic structures of tests are still defined in the grammar. Such a strategy avoids unnecessary redundant tests and complicated constraint solving for automatic test generation.

In this paper, we further extend the concept of a symbolic grammar by allowing a symbolic terminal to be carried over as a parameter through a production rule. This is an important and useful extension for software testing and test generation because symbolic grammars provide users great power and flexibility to express test cases, and the context-sensitive constraints defined over symbolic terminals can be collected and handled at runtime.

The LCT_{SG} System is an improved version of our previous work [6, 7], where we introduced a logic-based framework to implement an automated LSC simulator with practical user controls, and to test the consistency of an LSC specification with a set of enumerative test inputs. The LCT_{SG} system utilizes a memoized depth-first testing strategy to simulate how a reactive system model in LSCs behaves, in term of tree traversal, with infinite continuous test inputs. The computational tree is adapted in such a way that symbolic terminals are handled dynamically during the scenario

[1] The subscript *SG* means Symbolic Grammar.

simulation. Constraints on symbolic terminals may be collected and processed along the simulation to properly decompose their symbolic domains for branched testing.

The LCT_{SG} system further produces either a state transition graph or a failure trace, with branched testing and symbolic domain decomposition, if necessary, to justify the consistency testing results. Test generation using symbolic grammars provides an automatic and systematic way to produce a set of test cases. Combined with the state transition graph, symbolic grammars can be potentially evolved for refined test generation. Examples will be given to illustrate how grammar evolution is possible for refined test generation.

The paper is structured as follows. Section 2 briefly introduces the syntax and informal semantics of LSCs through a web order example. Section 3 presents the architecture of the LCT_{SG} system, a Java interface, logic programming backend LSC consistency testing system. Section 4 introduces the concept of symbolic grammars in automatic test generation, and shows how the symbolic terminals are defined and passed as a parameter of production rules for supporting context sensitivity. Section 5 illustrates how a running LSC reacts to the test events using a computational tree, and how symbolic terminals are processed through a computational path with domain decomposition and context sensitivity. Section 6 shows the supporting evidence for LSC consistency testing and simulation, and how the evidence can be further used to refine test generation. Finally, conclusions are given in Section 7.

2 Live Sequence Chart (LSC)

Live sequence charts (LSCs) [4, 11] have been introduced as an inter-object scenario-based specification and visual programming language for reactive systems. The language extends traditional sequence charts, typically sequence charts in UML [20] and message sequence charts (MSCs) [12], by enabling mandatory behaviors and forbidden behaviors specification. With mandatory behaviors, LSCs are able to specify necessary system behaviors in a more semantically precise manner instead of only posing weak partial order restriction on possible behaviors as in sequence charts. On the other hand, a forbidden behavior specification further enriches the LSCs with self-contained efficient scenario testing capability; it allows the LSCs to specify the failure scenarios and to locate those failures at runtime by checking reachability instead of testing failure conditions explicitly and blindly at every running step.

The language introduces a *universal* chart to specify mandatory behaviors in a more semantically precise manner. A universal chart is used to specify a scenario-based *if-then* rule, which applies to all possible system runs. A universal chart typically contains a *prechart*, denoted by a top dashed hexagon, and a *main chart*, denoted by a solid rectangle right below a prechart; if the prechart is satisfied, then the system is forced to satisfy or run the defined scenario in the main chart. The capability of specifying such mandatory behaviors upgrades the LSCs from a formal specification tool to an achievable dream of becoming a scenario-based programming language [8], because mandatory behaviors precisely tell what to expect at the system runtime.

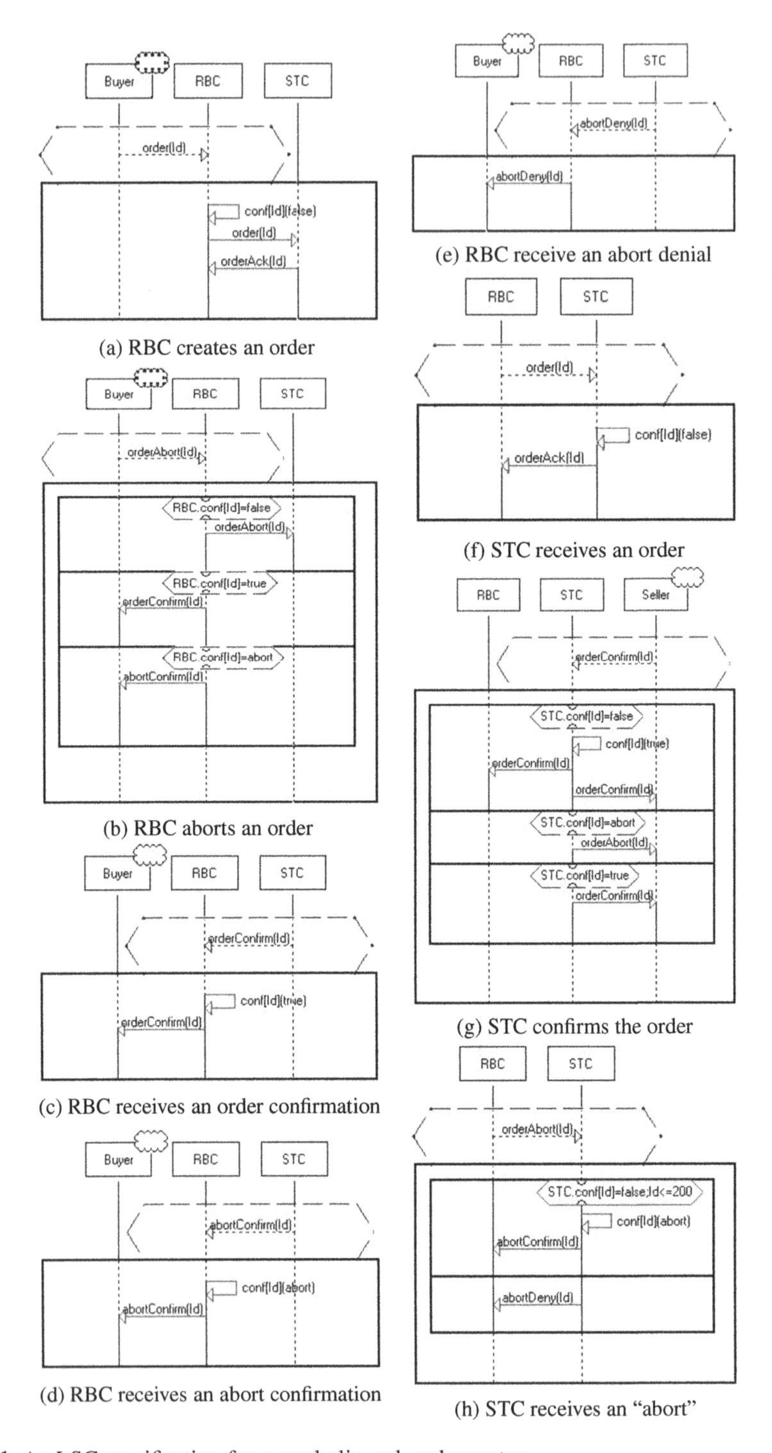

Fig. 1 An LSC specification for a symbolic web order system

2.1 An Indexed Web Order Protocol

We adopt an indexed web order protocol as an example to illustrate how LSCs are used to model a reactive system. Figure 1 shows eight universal LSCs for modeling a web order protocol between a really big corporation (RBC) and a small T-shirt company (STC). There are *two* types of orders created daily between RBC and STC, and each order has a unique numeric index *ID* from 1 to 300. One is a regular order with $1 \leq ID \leq 200$, in which the buyer of RBC could abort as long as the seller of STC has not confirmed the order; the other is a customized order with $200 < ID \leq 300$, in which once placed by the buyer, no abort action will be accepted. The order index will be reset to 1 daily, and increased by 1 automatically for a new order. For each order with an index *ID*, both RBC and STC has a state variable, *conf[ID]*, with the three possible values {*false, true, abort*} denoting whether the order has been initiated, confirmed or aborted, respectively; and the initial values of both state variables are *true*. We assume that there will be at most one active order at any time; that is, a new order can be initiated only if all previous orders have either been confirmed or aborted.

LSCs (a)(b)(c)(d)(e) illustrate scenarios for the RBC, and the rest LSCs describe the scenarios for the STC. Buyer and Seller are the only external objects, denoted by waved clouds, which send the external/user events, (*order(Id)*, *orderAbort(Id)*, and *orderConfirm(Id)*), to the system model. The rest of events are called *internal events*. Chart (a) shows if a buyer creates an order with an index *ID*, then the order is initiated in RBC by setting *RBC.conf[Id] false*, forwarding an *order* message to STC, and waiting for the acknowledgment. Chart (b) shows if the buyer wants to abort the order, and at that point if the order has been initiated but not confirmed yet, an abort message will be forwarded to STC, or otherwise, an appropriate message on the order status will be issued to the Buyer. The three stacked rectangles within the main chart is a multi-way selection construct, where each hexagon contains a condition. Chart (h) says that if STC receives an abort request, and at that point if the order has not been aborted or confirmed yet, and it is a regular order with $Id \leq 200$, it will set *STC.conf[Id]* to *abort* and send a confirmation message to RBC; otherwise, an abort denial will be issued to the RBC. Explanation for other LSCs are quite straightforward, therefore omitted here.

3 LCT$_{SG}$: LSC Consistency Testing with Symbolic Grammar

The architecture of our LCT$_{SG}$ system, illustrated in Figure 2, shows an overview of the interaction between the scenario-based programming tool, Play-Engine [11], and a logic-based LSC simulator through a Java frontend interface.

The Play-Engine [11] is a supporting tool for creating LSCs by playing their reactive behaviors directly from the system's graphical user interface. The LCT$_{SG}$ system complements the Play-Engine users [11] with an executable platform for automatic simulation, debugging, consistency testing capabilities, which are critical to the early designs of trustworthy system development processes. The LCT$_{SG}$

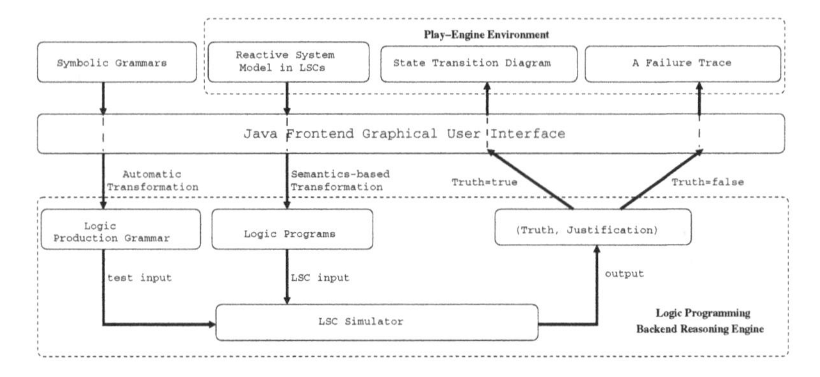

Fig. 2 The architecture of LCT_{SG}

system is a hybrid with a Java interface and a logic-based backend LSC simulator using SWI-Prolog engine [23], where the communication between Java and Prolog utilizes InterProlog's Java API [5], calling any Prolog goal through a *PrologEngine* object.

The LCT_{SG} system, given an LSC-based model of a reactive system and a symbolic grammar based specification of external event sequences, tests the consistency of the LCS specification by simulating the running LSCs on each given external event sequence. Our LSC simulator returns a truth value as well as its justification, which provides simulation evidences so that PLAY-engine users can easily follow the evidences to re-establish the simulation scenarios in an interactive way. The LCT_{SG} system further generates a state transition diagram, if the LSCs are consistent, illustrating the whole system behaviors, property satisfiability, as well as necessary domain decomposition for symbolic terminals. On the other hand, a graphical failure trace is constructed for the inconsistency evidence. Guided by justification results, LSC users can easily trace any particular system state or inconsistent scenario in the Play-engine.

In this paper, we focus on the introduction of a symbolic grammar for automatic test generation, how symbolic terminals affect the LSC testing, and how the symbolic grammar could be potentially evolved for refined test generation. Both our LCT_{SG} system and the PLAY-engine [11] assume an input-enabled concurrency model, where no two external events occur at the same time, and that before processing a next external event, the simulated system always reaches a stable state, where all enabled internal events have been processed.

4 Symbolic Grammars

A symbolic grammar $G_s = (V, \Sigma, \Sigma_s, W, P)$, where W is the main variable in V, extends the standard notation of a context-free grammar by the following additions:

- It introduces symbolic terminals Σ_s and domains in the grammar to hide the complexity of large redundant inputs which may have common syntactic pattern and similar expected behaviors. A symbolic terminal in Σ_s is represented in the form of $\{\alpha : D\}$, denoting a symbolic terminal name α with its associated finite domain D. The finite domain for a symbolic terminal can be represented as a list of ordered items in a form of either $Lower..Upper$ or a single value. For example, $[5..50, 60, 100..150]$ is a valid domain for a symbolic terminal.
- The latter occurrences of a same symbolic terminal $\{\alpha : D\}$ in the same production rule may be simply specified in the form of $\{\alpha\}$ omitting the domain. It raises a big difference between the proposed symbolic grammar and traditional CFG that a same symbolic terminal and its finite domain may be carried over within a production rule at runtime.
- A symbolic terminal $\{\alpha : D\} \in \Sigma_s$ can be used as a traditional terminal (e.g., $\{\alpha : D\} \in \Sigma$), a parameter of a terminal (e.g., $t(\{\alpha : D\}) \in \Sigma$), or even a parameter of a non-terminal (e.g., $X(\{\alpha : D\}) \in V$). It is very flexible to use a symbolic terminal and its associated domain to shadow the complexity of different inputs which share common syntactic structures. A symbolic terminal is allowed to pass among production rules to carry over the context-sensitivity, which is very useful to collect and solve those related constraints during the system simulation. Note that only symbolic terminals are allowed to be passed as parameters of terminals or non-terminals.

Example 1. *Consider the following symbolic grammar* $G = (V, \Sigma, \Sigma_s, W, P)$ *as an input of* LCT_{SG} *testing the LSCs shown in Figure 1, where the non-terminal set* $V = \{W, X, Y(\{id\})\}$*, the terminal set*

$$\Sigma = \{order(\{I\}), orderConfirm(\{I\}), orderAbort(\{I\})\},$$

the symbolic terminal set $\Sigma_s = \{\{id : [1..300]\}\}$*, and P is a set of symbolic production rules defined as follows:*

$$W \rightarrow \lambda \mid XW \tag{1}$$

$$X \rightarrow order(\{id : [1..300]\})\, Y(\{id\}) \tag{2}$$

$$Y(\{I\}) \rightarrow orderConfirm(\{I\}) \tag{3}$$

$$Y(\{I\}) \rightarrow orderAbort(\{I\})\, orderConfirm(\{I\}) \tag{4}$$

This symbolic grammar would generate continuous inputs of orders, and each generated order is followed by either an order confirmation directly, or a combination of an abort request and then an order confirmation. A symbolic terminal $\{id : [1..300]\}$ is used to replace the production of order IDs from 1 to 300. Two occurrences of {id} in a same production rule (e.g., rule (2)) share the same domain, and are expected to share the same instantiation as well during the LSC simulation; that is, any domain decomposition or instantiation of {id} in the event of *order* may be carried over to the same {id} in the following event. The domain decomposition or instantiation of {id} can be further passed into the rules (3) and (4) as an argument of the non-terminal Y.

The automatic test generation is done by simulating the leftmost derivation from its main variable W. Consider the grammar defined in Example 1. If we ignore the symbolic terminal, a valid leftmost derivation for the grammar would be:

$$\begin{aligned} W \Rightarrow XW &\Rightarrow order(id:[1..300])\ Y(id)\ W \\ &\Rightarrow order(id:[1..300])\ orderConfirm(id)\ W \\ &\Rightarrow order(id:[1..300])\ orderConfirm(id) \end{aligned}$$

During the LSC consistency testing, a symbolic terminal may be instantiated as a concrete value of its domain for a particular testing path, or its domain may be decomposed for branch testing during simulation. In general, such a symbolic grammar reduces several order-of-magnitude number of inputs to be enumerated. Nevertheless, the generated test inputs with symbolic terminals have enough nondeterminism to symbolically explore all running paths of the LSC simulation. Users could argue that a symbolic grammar does not expand the full syntactic structures of inputs due to the involvement of domain representations. However, the uses of symbolic terminals are perfect for system testing because it provides users great flexibility to shadow certain trivial details and reduce unnecessary testing complexity; at the same time, the introduction of symbolic terminals raises limited complexity of constraint solving due to the fact that symbolic terminals are only allowed at the leaf level of grammar.

5 Consistency Testing Using Symbolic Grammars

The LCT_{SG} contains a running LSC simulator, which takes inputs an LSC specification, *Ls* and a symbolic grammar G, to check whether the running *Ls* will react consistently to any external event sequence $w \in L(G)$. The computational semantics of the LSC simulator extends the operational semantics defined in [11] with the consideration of system object states, nondeterminism, and continuous environment reaction.

5.1 Super States

Similar to the super-step phase mentioned in the PLAY-engine [11], we consider super states only during the LSC simulation. Given an external event, the LSC simulator continuously executes the steps associated with any enabled internal events until the running LSCs reach a *stable* state where no further internal or hidden events can be carried out. A *super state*, defined as $\langle Q, RL, B\rangle$, is either an initial state or a stable state after executing an external event, where Q is a set of system object states, *RL* is the set of currently running copies of LSCs, and B indicates by *True* or *False* whether the state is a violating one. Let S_s denote a set of all super states and Σ_c a set of concrete external events without involving any symbolic terminals. We introduce a notation

$$\nabla : S_s \times \Sigma_c \to S_s^+$$

to denote the super-step transition in [11], which takes a super state and a concrete external event as inputs, and returns a set of all possible super states. Multiple next super states are possible because the enabled internal events may contain interleaving ones, which can be executed in a nondeterministic order.

5.2 *Operational Semantics*

We introduce a new succinct notation for describing the successive configurations of the LSC simulator given the input of a symbolic grammar $G = (V, \Sigma, \Sigma_s, W, P)$ denoting a set of external event sequences. A five-tuple

$$(Q, D, U, RL, B),$$

is introduced to describe the *running state* for the LSC simulator, where Q is a set of current object states defined in the system, U is the unprocessed part of the symbolic grammar in a sentential form, D is a set of symbolic terminals, associated with their respective domains, which are currently used in Q, RL is a set of current running/activating LSCs, and a boolean variable B indicates by *True* or *False* whether the state is a violating/inconsistent one. The running state completely determines all the possible ways in which the LSC simulator can proceed, where the symbolic terminal set D maintains the context sensitive information during the simulation.

A *valuation* θ_D for a symbolic terminal set D is an assignment of values to each symbolic terminal in D from their associated domains; for any notation or expression e, $e\theta_D$ is obtained by replacing each symbolic terminal in e by its corresponding value specified in the valuation θ_D. It needs to be clarified here that the five-tuple running state, possibly involving symbolic terminals, is introduced to configure our LSC simulation, while the super state, without symbolic terminals, is used to describe the operational semantics related to the PLAY-engine. The operational semantics of our LSC simulator, in terms of running states, moves and a PLAY-tree, is defined on the top of the underlying super states.

Definition 1 (Move). A *move* from one running state to another, denoted by the symbol $\vdash$, could be one of the following cases:

- if $a \in \Sigma$ is a *symbolic external event* (a terminal possibly involving symbolic terminals), and $U \in \{\Sigma \cup V\}^*$,

  $$(Q_1, D_1, aU, RL_1, B_1) \vdash (Q_2, D_2, U, RL_2, B_2)$$

  is valid if

  $$\forall \theta_{D_2}, \ (Q_2, RL_2, B_2)\theta_{D_2} \in \nabla((Q_1, RL_1, B_1)\theta_{D_2}, a\theta_{D_2}).$$

  Such a move is called a *symbolic terminal move*.

- if $A \in V$ is a nonterminal variable,

$$(Q, D_1, AU, RL, B) \vdash (Q, D_2, E_1 \cdots E_n U, RL, B)$$

is valid if $A \rightarrow E_1 \cdots E_n$ is a production rule in P, and $D_2 = D_1 \cup \{\text{any new symbolic terminals in } E_1 \cdots E_n\}$. Such a move is called a *nonterminal move*.

The validity of a symbolic terminal move is defined as that for each concrete instantiation of symbolic terminals in D_2, it has a valid corresponding super-step move in the PLAY-engine. For each symbolic terminal $\beta \in D$, we use $D(\beta)$ to denote the domain of β in the symbolic terminal set D. Thus, in a symbolic terminal move, we have $D_2(\beta) \subseteq D_1(\beta)$ due to the possible branched testing or instantiation during the simulation. A symbolic terminal may be removed from D_1 after a symbolic terminal move if the terminal does not occur in the unprocessed grammar part U. Different symbolic terminal moves from a same running state are possible due to inherent nondeterminism which could be caused by interleaving messages in an LSC, or the nondeterminism caused by domain decomposition for a symbolic terminal.

A nonterminal move extends the leftmost nonterminal A with one of its matched production rule; D_2 may add new symbolic terminals involved in the production rule. Different nonterminal moves from a same state are also possible due to possible multiple production rules from the same variable to represent different message sequence composition.

A PLAY-tree is a derivation tree used to illustrate the continuous and alternative configurations of a running LSC engine, where each derivation in the PLAY-tree corresponds to a valid nonterminal move or symbolic terminal move. In a PLAY-tree, each parent-child edge represents a valid move. For clarity, we add labels for each edge. For an edge of a symbolic terminal move, the label is in form of `[a]`, where a is a symbolic external event triggering the move; for an edge of a nonterminal move, the label is a production rule $A \rightarrow E_1 \cdots E_n$, which triggers the move.

5.3 Testing the Web Order System

Consider the LSC simulation for testing the web order system in Figure 1 with an input of the symbolic grammar shown in Example 1. Figure 3 shows a PLAY-tree in the LCT$_{SG}$ system for the web order example, where the bold solid box and the bold dashed one denote a leaf node and a variant node[2], respectively. The object state of the web order system is simply represented by the values of both `RBC.conf[id]` and `STC.conf[id]`, where $id \in \Sigma_s$ is a symbolic terminal with an initial empty domain. The abbreviations from **S1** to **S3** represent three different object states in the PLAY-tree; and the abbreviations from **D1** to **D3** represent three different sets of symbolic terminals.

Note that in each node of the PLAY-tree shown in Figure 3, the set of current running LSCs is always $\emptyset$. That is because in this web order system example, (i) no

[2] A node is called a *variant* of another if both nodes have the same five-tuple running state except the renaming of symbolic terminals.

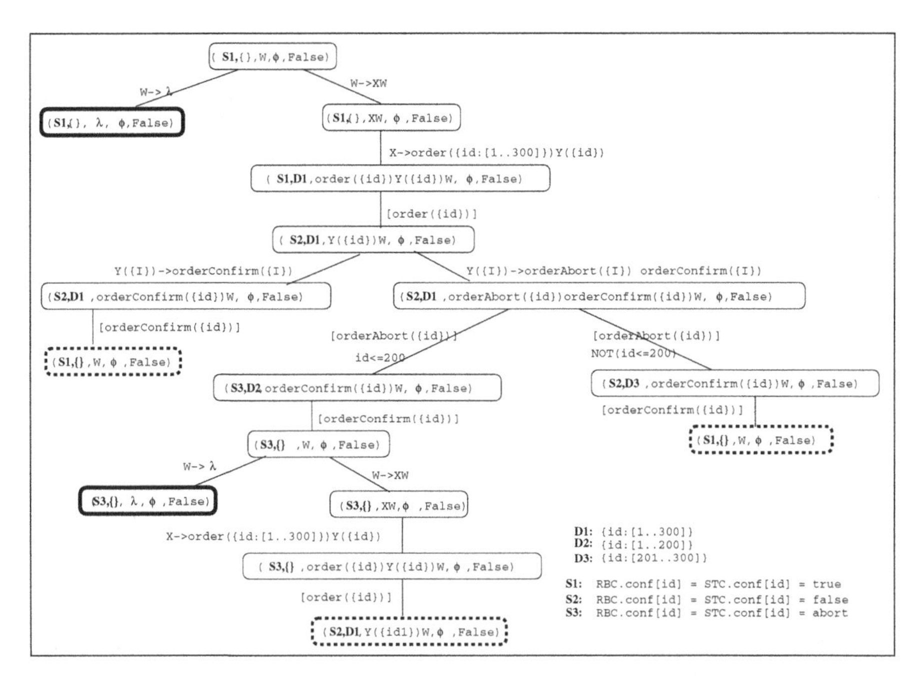

Fig. 3 A Play-Tree for Testing the Web Order System

LSCs are active initially; (ii) after receiving an external event and then processing all enabled internal events, all the activated LSCs will be complete before receiving a next external event. This is consistent to the simulation results in the super-step mode using the Play-engine. Also, we choose this example particularly to leave the details in [11] on how exactly the internal transition works. Now we highlight and explain important features in Figure 3 as follows:

Consistency Testing via a Tree Traversal. The consistency of an LSC model is defined in terms of a PLAY-tree. It is consistent if all the finite branches in the PLAY-tree are success branches; otherwise, the LSC model is inconsistent if there exists a failure branch in the PLAY-tree; and a prefix of failure trace can be easily located along the failure branch. Even though the symbolic grammar G may contain infinite sequences, for each finite sequence $w \in L(G)$, there is a *corresponding* finite branch in the PLAY-tree, along which if we concatenate all the labels of terminal moves from the root, the result will be the finite event sequence w. The LSCs of the web order system described in Figure 1 on the given symbolic grammar G is consistent because for each finite sequence $w \in L(G)$, there is a *corresponding* success branch in the PLAY-tree in Fig. 3.

Memoized Depth-First Testing Strategy. Given an LSC model, the consistency checking is a traversal of its PLAY-tree to see whether there exists any failure branch. Due to the recursion nature of the symbolic grammar G and non-determinism nature of LSCs, there may exist many infinite branches in the

PLAY-tree, or the finite branch could be any long. Therefore, neither depth-first nor breadth-first strategy is good enough for the completion of consistency check over the PLAY-tree. We apply a memoized depth-first search strategy for traversing a PLAY-tree from left to right, such that any variant nodes along the path seen later will not be explored again because its running behaviors would be same as the previous one. A memoized depth-first strategy is an extension of standard depth-first search where visited nodes along the path are recorded so that their later variant occurrences can be considered a cycle of moves, thus dramatically simplify the search process. The variant nodes are shown in the dashed bold boxes in Fig. 3, where renaming on symbolic terminals are allowed.

Leftmost Derivation with Context Sensitivity. Given a symbolic grammar G, the LCT_{SG} system produces every symbolic input sequence in $L(G)$ by simulating the leftmost derivation from its main variable W. During the derivation,

- if the leftmost symbol is a variable (e.g., the root node $(S1, \{\}, W, \emptyset, False)$), the system takes a non-terminal move by applying a matched production rule (e.g., $W \rightarrow XW$).
- if the leftmost symbol is a variable (e.g., $(S1, \{\}, XW, \emptyset, False)$), and the matched production rule contains a new symbolic terminal in its body ($X \rightarrow order(\{id : [1..300]\})Y(\{id\})$), the system adds *id* and its associated domain into the symbolic terminal set (see the move in Figure 3). The latter occurrences of the same *id* will apply their associated domain from the symbolic terminal set. The *id* and its associated domain will be removed from the symbolic terminal set once both $order(\{id\})$ and $Y(\{id\})$ are completely processed. Each symbolic terminal has a live scope during the derivation.
- if the leftmost symbol is a variable with a symbolic terminal as a parameter (e.g., $(S2, D1, Y(\{id\})W, \emptyset, False)$), the system takes a non-terminal move by applying a matched production rule (e.g., $Y(\{I\}) \rightarrow orderConfirm(\{I\})$), and the symbolic domain will be kept same as $D1$.
- if the leftmost one is a terminal with a symbolic terminal as a parameter (e.g., $(S2, D1, orderConfirm(\{id\})W, \emptyset, False)$), the system will find its corresponding domain from the set $D1$, and continue its simulation. Once $orderConfirm(\{id\})$ has been processed during the LSC simulation, the *id* will be removed from the symbolic domain.

In our web order example, we assume that there is at most one active order at any time, and every new order comes independently from the previous orders.

Domain Decomposition for Branched Testing. In the LCT_{SG} simulation, constraints on a symbolic terminal may be dynamically collected and processed to properly instantiate the symbolic terminal or decompose its domains for branched testing. As shown in the PLAY-tree of Figure 3, the symbolic terminal move from the node $(\mathbf{S2}, \mathbf{D1}, orderAbort(\{id\})orderConfirm(\{id\})W, \emptyset, False)$ split into two branches, where the symbolic domain of *id* is decomposed into two sub-domains [1..200] and [201..300]. The domain decomposition is caused by the related constraint "`id<=200`" defined in the LSC of Figure 1(h). Whether this constraint is satisfied or not leads to two different running states during the LSC simulation. Our LCT_{SG} simulator processes such a conditional constraint "`id<=200`", combined with the original domain $id : [1..300]$, to reduce the domain to $id : [1..200]$ for further simulation. Once the simulation is done on this branch, the LCT_{SG} automatically backtracks to this conditional point to explore the branch with its negative constraint "`id>200`". This is a very important feature of using symbolic terminals, avoiding unnecessary redundancy for testing each value in its domain, yet still maintaining enough flexibility to explore non-determinism and alternatives.

6 Grammar Evolution Using Justification

Given an LSC specification and a symbolic grammar, the LSC system returns a truth value of consistency as well as its justification [18], which provides evidences so that users can easily follow the evidences to re-establish the simulation scenarios. If the consistency is false, it returns a failure trace as a negative justification; while if the result is true, it actually returns a state transition diagram (STD) as a positive justification, showing labeled transitions between states while processing each external event. The LSC uses the graph visualization, *Graphviz*, to generate STD diagrams automatically. Figure 4 shows a STD diagram as a positive justification for the consistency testing of the web order system. Each node in the diagram represents a system state including the status of system objects; each transition is labeled with an external event including symbolic terminals with associated domains. The transition from nowhere indicates the initial state; and the ε-transition corresponds to a system procedure where an symbolic terminal is removed from the symbolic terminal set.

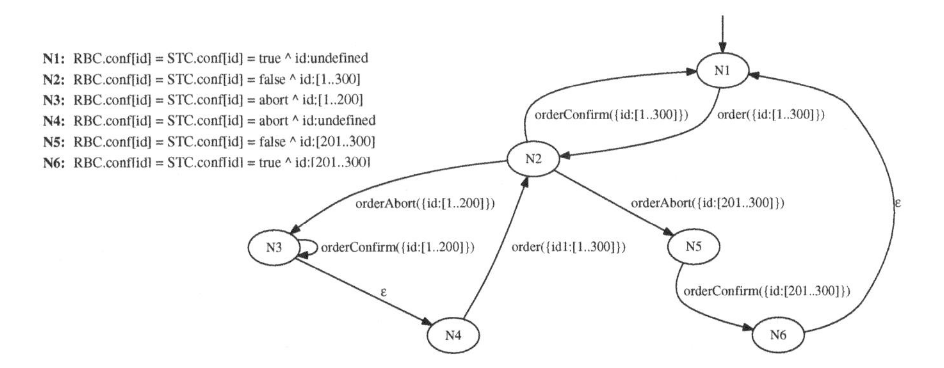

Fig. 4 Justification with State Transition Diagram

The STD diagram is essentially a projection of the Play-Tree in Figure 3 on the focus how the symbolic external events affect the web order system. The self-loop transition, $orderConfirm(id : [1..200])$ on state $N3$, tells that for any order

$id \in [1..200]$, if the order has been aborted, the following *orderConfirm* message is void. The diagram also shows how the symbolic domain may be passed along the model-based simulation and may be partitioned for branched situational testing. The *orderAbort* transitions from state $N2$ are branched due to the domain partition, where the left branch corresponds to the satisfaction of the system constraint $id \leq 200$.

The LCT_{SG} introduces a symbolic grammar for automatic test generation. The STD diagram of justification may be used to evolve the symbolic grammar for refined test generation. Consider the symbolic grammar in Example 1 and the STD diagram in Figure 4. Three types of evolution could be potentially applied to refine the symbolic grammar.

- The branched testing with domain partition in the STD diagram suggests that the test generation on *orderAbort* may be separated as follows for different situational cases:

$$Y(\{I\}) \rightarrow \ orderAbort(\{I : [1..200]\})\ orderConfirm(\{I\})$$
$$Y(\{I\}) \rightarrow \ orderAbort(\{I : [201..300]\})\ orderConfirm(\{I\})$$

- The self-loop transition on State $N3$ denotes void external events, which should be removed from the test generation for the reason of efficiency. Thus, The production for Y({I}) can be further refined to

$$Y(\{I\}) \rightarrow \ orderAbort(\{I : [1..200]\})$$
$$Y(\{I\}) \rightarrow \ orderAbort(\{I : [201..300]\})\ orderConfirm(\{I\})$$

- Based on the STD diagram, users can specify state properties serving as extra test generation directives. For example, users would like to test those system scenarios where the order will be eventually confirmed, that is,

$$RBC.conf[id] = RBC.conf[id] = true.$$

  Thus, only the following production shall be included in the refined grammar:

$$Y(\{I\}) \rightarrow \ orderAbort(\{I : [201..300]\})\ orderConfirm(\{I\})$$

Not only does the LCT_{SG} provide an executable environment for formal simulation and consistency testing, it also presents a new systematic way for model-based test generation. The LCT_{SG} system accepts two inputs, the LSCs as a formal model of the software under test and a preliminary symbolic grammar as test generation directives, and provides mechanisms to evolve the system grammar for refined test generation. The preliminary symbolic grammar may describe the test cases independently from the system requirement and design, while an evolved symbolic grammar takes an executable system model into consideration for refined test case generation. Detailed formalization on grammar evolution will be given in our future work.

7 Conclusions

We presented the LCT_{SG}, a Java interface and a logic-programming backend LSC testing and simulation system. The LCT_{SG} takes LSCs and symbolic grammars as inputs and performs an automated LSC simulation for consistency testing. A test generation using symbolic grammars avoids both unnecessary redundant tests and expensive constraint solving techniques for automatic test generation. Our LCT_{SG} complements the Play-Engine system with an executable platform for automatic simulation, debugging, and consistency testing capabilities. It provides great power in a formal testing environment to support large-scale system development.

The LCT_{SG} system further provides a state transition diagram to justify the consistency checking results and evolve the grammatical input for refined test generation. We believe that an automatic LSC simulator as well as debugging, consistency checking capabilities, automatic evolving test generation will play important roles on designing trustworthy reactive distributed software processes.

References

1. Bartsch, K., Robey, M., Ivins, J., Lam, C.: Consistency checking between use case scenarios and uml sequence diagrams. In: IASTED Conference on Software Engineering, pp. 581–589 (2004)
2. Bontemps, Y., Heymans, P.: Turning High-Level Live Sequence Charts into Automata. In: Pacholski, L., Tiuryn, J. (eds.) CSL 1994. LNCS, vol. 933, pp. 456–470. Springer, Heidelberg (1995)
3. Clarke, L.A.: A system to generate test data and symbolically execute programs. IEEE Transaction on Software Engineering 2, 215–222 (1976)
4. Damm, W., Harel, D.: Lscs: Breathing life into message sequence charts. In: Proceedings of 3rd IFIP Int. Conf. on Formal Methods for Open Object-based Distributed Systems, pp. 293–312 (1999)
5. Declarativa. Interprolog 2.1.2: a java front-end and enhancement for prolog (2010), `http://www.declarativa.com/interprolog`
6. Guo, H.-F., Zheng, W., Subramaniam, M.: Consistency checking for LSC specifications. In: 3rd IEEE International Symposium on Theoretical Aspects of Software Engineering, TASE 2009 (July 2009)
7. Guo, H.-F., Zheng, W., Subramaniam, M.: L2c2: Logic-based LSC consistency checking. In: 11th International ACM SIGPLAN Symposium on Principles and Practice of Declarative Programming, PPDP (September 2009)
8. Harel, D.: From Play-In Scenarios to Code: An Achievable Dream. In: Maibaum, T. (ed.) FASE 2000. LNCS, vol. 1783, pp. 22–34. Springer, Heidelberg (2000)
9. Harel, D., Kugler, H.: Synthesizing state-based object systems from LSC specifications. Int. Journal of Foundations of Computer Science 13(1), 5–51 (2002)
10. Harel, D., Maoz, S., Segall, I.: Some results on the expressive power and complexity of LSCs. In: Pillars of Computer Science, pp. 351–366 (2008)
11. Harel, D., Marelly, R.: Come, Let's Play: Scenario-Based Programming Using LSCs and the Play-Engine. Springer (2003)
12. ITU-T. Message sequence chart (MSC). Z.120 ITU-T Recommendation (1996)
13. King, J.C.: Symbolic execution and program testing. Commun. ACM 19, 385–394 (1976)

14. Klose, J., Toben, T., Westphal, B., Wittke, H.: Check It Out: On the Efficient Formal Verification of Live Sequence Charts. In: Ball, T., Jones, R.B. (eds.) CAV 2006. LNCS, vol. 4144, pp. 219–233. Springer, Heidelberg (2006)
15. Kugler, H., Harel, D., Pnueli, A., Lu, Y., Bontemps, Y.: Temporal Logic for Scenario-Based Specifications. In: Halbwachs, N., Zuck, L.D. (eds.) TACAS 2005. LNCS, vol. 3440, pp. 445–460. Springer, Heidelberg (2005)
16. Majumdar, R., Xu, R.-G.: Directed test generation using symbolic grammars. In: The ACM SIGSOFT Symposium on the Foundations of Software Engineering: Companion Papers, pp. 553–556 (2007)
17. Maurer, P.M., Maurer, P.M.: Generating test data with enhanced context-free grammars. IEEE Software 7, 50–55 (1990)
18. Pemmasani, G., Guo, H.-F., Dong, Y., Ramakrishnan, C., Ramakrishnan, I.: Online justification for tabled logic programs. In: The 7th Int. Symposium on Functional and Logic Programming (2004)
19. OMG. Unified modeling languages superstructure specification, v2.0. The Object Management Group (2005), `http://www.uml.org/`
20. OMG. Documentation of the unified modeling language. The Object Management Group (2009), `http://www.omg.org`
21. Sirer, E.G., Bershad, B.N.: Using production grammars in software testing. In: Second Conference on Domain Specific Languages, pp. 1–13 (1999)
22. Sun, J., Dong, J.S.: Model checking live sequence charts. In: The 10th IEEE Int. Conf. on Engineering of Complex Computer Systems (2005)
23. SWI-Prolog. SWI-prolog's home (2010), `http://www.swi-prolog.org`
24. Toben, T., Westphal, B.: On the expressive power of LSCs. In: The 32nd Conf. on Current Trends in Theory and Practice of Computer Science, pp. 33–43 (2006)
25. Visser, W., Păsăreanu, C.S., Khurshid, S.: Test input generation with java pathfinder. SIGSOFT Softw. Eng. Notes 29, 97–107 (2004)

On the Need of New Approaches for the Novel Problem of Long-Term Prediction over Multi-dimensional Data

Rui Henriques and Cláudia Antunes

Abstract. Mining evolving behavior over multi-dimensional structures is increasingly critical for planning tasks. On one hand, well-studied techniques to mine temporal structures are hardly applicable to multi-dimensional data. This is a result of the arbitrary-high temporal sparsity of these structures and of their attribute-multiplicity. On the other hand, multi-label classification over denormalized data do not consider temporal dependencies among attributes.

This work reviews the problem of long-term classification over multi-dimensional structures to solve planning tasks. For this purpose, firstly, it presents an essential formalization and evaluation method for this novel problem. Finally, it extensively overviews potential relevant contributions from different research streams.

1 Introduction

New planning opportunities are increasingly triggered by the growing amount, completeness and precision of temporal data. The integration of data in multi-dimensional structures have been enabled through the world-wide adoption of data warehouses. For this setting, the study of long-term prediction in evolving contexts can increasingly provide additional value [6][34].

Applications may range from clinical prevention to several planning tasks as in retail, educational, commercial, financial and social security domains [26][6]. An example may be the long-term planning of hospital resources based on underlying healthcare needs (for instance, seen as the need of a patient get a specific treatment within upcoming years).

Rui Henriques · Cláudia Antunes
DEI, IST–UTL, Portugal
e-mail: rmch@ist.utl.pt, claudia.antunes@ist.utl.pt

R. Lee (Ed.): Computer and Information Science 2012, SCI 429, pp. 121–138.
springerlink.com

The mining of temporal dynamics using multistep-ahead classifiers has been mainly applied to temporal and sequential structures [48][11]. In practice, this body of knowledge is hardly applicable to multi-dimensional data structures. Although mappings between these structures exist, the resulting temporal event-sparsity and attribute-multiplicity claim for new research. Additionally, although multi-label classifiers can be adopted by denormalizing multi-dimensional into tabular structures, they fail to deal with temporal dependencies.

Further challenges of long-term prediction in evolving contexts include the ability to deal with different time scales [1][7], advanced temporal rules [4] and knowledge-based constraints for an accurate and efficient long-term learning with minimum domain-specific noise [3].

When considering, for example, an hospital planning task, multi-dimensional structures are centered in health-records (the fact) grouped by a tuple-aggregator dimension, usually by patient. Health records may track measures of interest related to monitoring, diagnostic, prescription or treatment dimensions. This structure is not in the form of a temporal structure for the ready application of existing well-studied predictors. Additional challenges include the arbitrary sparse nature of patient visits and of measure recordings.

The document is structured as follows. Section 2 discusses the novelty of this problem and introduces a case to illustrate its significance. Section 3 formulates the problem of long-term prediction over multi-dimensional structures. Section 4 discusses the key aspects for its correct assessment. Finally, an overview of the relevant work in long-term prediction is synthesized and existing contributions plugged as potential principles to address the problem.

2 Why a New Long-Term Prediction Formulation?

Consider a training dataset following a simple multi-dimensional structure, a star scheme centered in one fact that track events unevenly spread across time. In healthcare, a health record flexibly captures measures related to standardized dimensions as laboratory results, prescriptions, treatments and diagnostics. Health records can be used to learn a multi-step-ahead learner that classifies a measure across multiple periods for planning tasks as, for instance, the hospitalization length of a patient in the upcoming months. Lower-level planning tasks (as the required number of a specific treatment), or personalized tasks (as the prediction of evolving physiological states) can be additionally consider.

Despite the relevance of this problem, to the best of our knowledge there is not yet a dedicated research stream for its solution. Fig.1 gathers the existing research streams that can provide important contributions to the development of novel approaches for long-term prediction. These contributions may be applied *as-is* or through mappings into tabular and temporal structures, as illustrated in Fig.2. Next subsections synthesize their limitations.

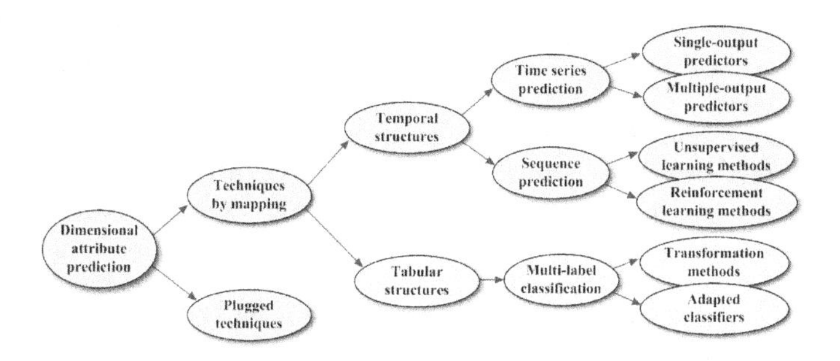

Fig. 1 Key areas for the definition of principles across different settings

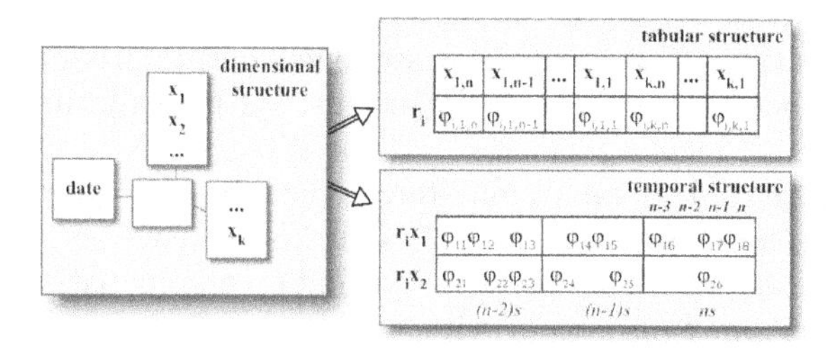

Fig. 2 Structural mappings for an adapted multistep-ahead classification

2.1 *Limitations of Multi-label Classifiers*

A simple procedure to deal with prediction over multi-dimensional structures is to denormalize these structures into a plain tabular structure and to apply a multi-label classifier.

A first set of challenges appear as multi-label classifiers were developed with a different purpose – categorization and multi-parameter diagnosis. *First*, multi-label classifiers are not prone to label ordinal attributes, but to label disjoint binary attributes. Labeling of nominal attributes requires the mutual exclusive labeling of binary attributes. *Second*, attributes are considered to be independent. Temporal dependencies are not captured in the learning process.

Additionally, two core and severe problems arise when adopting tabular structures. *First*, when capturing each event occurrence as an attribute, the size of the table may grow dramatically, which can significantly reduce the efficiency of the learning process and the accuracy of the classifier. *Second*, its viability strongly depends on the ability to capture the temporal dependencies both among domain attributes and across the periods under prediction.

2.2 Limitations of Sequence and Time Series Predictors

Given a training dataset with series composed by $n+h$ observations, *multi-step-ahead prediction over univariate series* is the task of learning a model to predict the h next observations of a n-length series, where $h>1$ is the *prediction horizon.*

These series can either be sequences, as a genetic arrangement, or time series, as a physiological signal. Research streams on these areas are important if we want to work with one measure-of-interest. To deal with the multiple measures captured in a multi-dimensional structures, one can treat these measures as a multivariate sparse temporal structure. Let us assume that a mapping between a dimensional and a temporal structure is possible. In this way, the problem would be the prediction of multiple periods based on two types of input: sparse multivariate time sequences derived from fact measures, and static attributes derived from dimensions. With this formulation, three challenges arise.

The *first challenge* is to adapt long-term predictors to deal with arbitrary-high sparse time sequences. The rate of events' occurrence per fact may vary significantly across time. This happens within and among patients in health-care systems, but may also occur with banking applications, genetic mutations or almost every dynamic daily-life system. Structural sparsity results from the alignment of events across time points. Sequential predictors can solve this problem as they are focused on causalities [40]. However, since they do not account for temporal distances among events, they are non-expressive candidates.

The *second challenge* is to enrich long-term prediction in order to deal with multivariate time sequences. For instance, determining if a patient is hospitalized across multiple periods is conditionally dependent on the patient clinical history composed by multiple attributes as optional prescriptions, symptoms, exams, treatments and diagnosis. These measures, if seen as multivariate time sequence, can also be subjected to prediction.

The *third challenge* is to perform long-term prediction in evolving contexts. The relevance of understanding evolutionary behavior in planning problems through predictive rules is discussed in [26][22]. Both predictors with propensity towards over-fitting or extensive smoothing fail to capture exclusive behavior of an instance under prediction. Predictors can either scan large or local periods of a fact measure depending if the measure is considered to be stationary or non-stationary. In the first case, cyclic behavior can be considered [26]. In evolving contexts, temporal partitions with a significant number of occurrences must be selected. Challenges on how to deal with evolving behavior depend on the adopted approach. In the second case, the understanding of evolving and calendric behavior may be used to complement the supervised learning.

2.3 Illustration

The challenges of mining temporal aspects in dimensional data centered in health records are synthesized in Table 1.

Table 1 Requirements for long-term prediction over health records

Healthcare data challenges	Requirement
Health records may define multiple measures of interest;	Strategies to deal with multivariate structures
The number of health records can be significantly high and its flexible nature may hamper the learning;	Background knowledge guidance to avoid efficiency and domain-noise problems
Health records are irregularly collected due to an uneven schedule of visits or measurements made;	Methods to deal with missing values and event sparsity
Health record sampling grid varies both among and within patients;	Efficient structural operations for record alignment and time partitioning
Different measures can be recorded at distinct time scales;	Calendric-mining and aggregation techniques to deal with the different sampling rate of health records
Evolving patterns, as the progress of a disorder or a reaction to a prescription, are spread across many non-relevant records;	Convolutional memory techniques and pattern-based learning ability to detect evolving health trends

3 Problem Formulation

This section formalizes the target problem. For this task, a formal review of the underlying concepts is introduced.

3.1 Underlying Definitions

Def. 1. Consider a dataset of training instances, $D = \{x_1, ...x_n\}$, of the form $x_i = (a, \mathbf{y})$, where $a = \{a_1 \in A_1, .., a_m \in A_m\}$ is a set of input attributes and $\mathbf{y} = (y_1, .., y_h) \in Y^h$ is a vector of either numeric or categorical symbols, where $h > 1$ is the horizon of prediction. Given a training dataset, the task of *long-term prediction* is to learn a mapping model $M : A \to Y^h$ that accurately and efficiently predicts y based on a particular x, i.e. $y{=}M(x)$.

Given a training dataset D, the task of *long-term prediction over tabular data* is to learn a model $M : A \to Y^h$, where the domain is described by a set of alphabets, $A = \{\Sigma_1, .., \Sigma_m\}$.

Although this problem is similar to a multi-label classification problem, its definition explicitly considers conditional temporal-dependency among $\mathbf{y}$ symbols and includes ordinal attributes.

Def. 2. Let Σ be an alphabet of symbols σ, and $\tau \in \mathbb{R}$ be the sample interval of a series of equally-distant time points, $\{\theta_i \mid \theta_i = \tau_0 + i\tau; i \in \mathbb{N}\}$. A *sequence* s is a vector of symbols

$s=(\varphi_1,..,\varphi_n)$, with $\varphi_i=[\varphi_{i,1},..,\varphi_{i,d}] \in (\mathbb{R}|\Sigma_i)^d$.

A *time series* t with regard to θ_i, is given by

$t=\{(\varphi_i,\theta_i)|\varphi_i=[\varphi_{i,1},..,\varphi_{i,d}] \in (\mathbb{R}|\Sigma_i)^d, i=1,..,n\} \in \mathbb{T}^{n,d}$.

A *time sequence* w is a multi-set of events

$w=\{(\varphi_i,\theta_j)|\varphi_i=[\varphi_{i,1},..,\varphi_{i,d}]\in (\mathbb{R}|\Sigma_i)^d; i=1,..,n; j\in\mathbb{N}\}$.

s, t and w are univariate if $d=1$ and *multivariate* if $d>1$.

The domains of sequences is $\mathbb{S}^{n,d}$, time series is $\mathbb{T}^{n,d}$, and time sequences is $\mathbb{W}^{n,d}$, where n is the length and d the multivariate order.

Exemplifying, a univariate time series capturing monthly hospitalizations can be $\mathbf{y}=\{(0,\tau_1),(3\text{-}5,\tau_2),(>5,\tau_3),(2,\tau_4)\}$, with $y \in \mathbb{T}^{4,1}$. A multivariate time sequence capturing two physiological measures from blood tests can be $a=\{([2\ 21],\tau_2),([3\ 19],\tau_3),([2\ 20],\tau_5)\}$, with $a \in \mathbb{W}^{6,2}$.

Given a training dataset D, the well-studied task of *time series long-term prediction* problem is to learn a mapping model $M: A \rightarrow Y^h$, where $A=\mathbb{T}^{m,1}$, and A and Y values are either numeric or share the same alphabet Σ.

Given a training dataset D, the task of *long-term prediction over multivariate and sparse temporal structures* is to construct a mapping model $M: A \rightarrow Y^h$, where $A=\mathbb{W}^{m,d}$.

Def. 3. Given a training dataset D with tuples in the form of $(a=\{a_1\in A_1,.., a_m\in A_m\},\mathbf{y}\in Y^h)$, the task of *long-term prediction over multi-dimensional data* is to construct a mapping model $M: \{A_1,..,A_m\} \rightarrow Y^h$, where a attributes are either a symbol or a time sequence of l-length and multivariate d-order ($X_i=\Sigma \mid \mathbb{W}^{l,d}$), $\mathbf{y}$ is a vector of h symbols, and $h>1$.

Exemplifying, an instance, $(\{x_1,x_2,x_3\},y)$, can represent a patient, where x_1 is his age, x_2 and x_3 are two multivariate time sequences capturing measures from blood and urine tests, and $\mathbf{y}$ is his number of hospitalizations across different periods. The goal is to learn a model, based on a training dataset, to predict multi-period hospitalizations $\mathbf{y}$ for a patient based on health-related data x.

3.2 Long-Term Prediction over Dimensional Data

In order to understand how to derive x from a multi-dimensional dataset, some concepts are formalized below.

A *multi-dimensional data structure*, $\{(\cup_i\{Dim_i\}) \cup Fact\}$, is defined by a set of dimensions Dim_i and one fact $Fact$. Each dimension contains a primary key, a set of attributes a, and no foreign keys. The fact contains one foreign key for each dimension and a set of measures $(b_1,..,b_d)$.

Two special dimensions, the *time* and *select* dimensions, need to be identified. The select dimension, the patient in the healthcare example, is used to group the multiple fact occurrences across time in n instances $(x_1,..,x_n)$ according to the primary keys in this dimension. The number of instances, n, is given by the number of these primary keys.

Given a multi-dimensional dataset D, its mapping in a set of instances of the form $(a_1,..,a_m)$ follows a three-stage process. *First*, using D select-dimension, the set of all fact occurrences is grouped in n instances $(x_1,..,x_n)$. *Second*, the set of fact occurrences for each instance is mapped into a multivariate time sequence, b=$\{(b_i,\theta_j)|\ \varphi_i$=$[b_{i,1},..,b_{i,d}]; i$=$1,..,n; j{\in}\mathbb{N}\}$, where the order d is the number of fact measures. *Third*, the attributes from dimensions are captured as one-valued attributes, a_{Dim_i}=$(a_{i,1},..,a_{i,|Dim_i|})$. After this three-step process the instances follow the form $(a_1,..,a_m)$, where a_i attribute is either derived from a dimension or a multivariate time sequence of l-length derived from a fact (a_i=$b \in \mathbb{W}^{l,d}$).

Illustrating, consider the multi-dimensional healthcare dataset $\{Dim_{patient}$= $\{id_2, A_{1,1}, A_{1,2}\}, Dim_{lab}$=$\{id_3, A_{2,1}\}, Dim_{time}, Fact_{hr}$=$\{fk_1, fk_2, fk_3, B_1, B_2\}\}$. An example of a retrieved patient tuple is $(a_{1,1}, a_{1,2}, a_{2,1}, \{([b_{1,1}\ b_{1,2}], \theta_1), ([b_{2,1}\ b_{2,2}], \theta_4), ([b_{3,1}\ b_{3,2}], \theta_5)\})$.

In real-world planning tasks, the training dataset may not be temporally compliant with the instance under prediction. Two strategies can be used in these cases: allowance of temporal shifts to the training tuples and/or transition from a pure supervised solution into an hybrid solution. For instance, if we consider health records between 2005 and 2011, and we want to predict the hospitalizations for a patient until 2014, the model can either rely on a 3-year temporal shift and/or the projection of cyclic and calendric patterns.

4 Evaluation

Long-term prediction requires different metrics than those used in traditional single-label classification. This section presents the set of metrics adopted in the literature, and proposes a roadmap to evaluate long-term predictors.

Predictor's efficiency is measured in terms of memory and time cost for both the training and testing stages.

The accuracy of a predictive model is the probability that the predictor correctly labels multiple time points, $P(\hat{y} = y)$. This probability is usually calculated over a train dataset using a 10-fold cross-validation scheme. If not, disclosure of the adopted sampling test technique (e.g. holdout, random sub-sampling, bootstrap) needs to be present. Accuracy can be employed using similarity or loss functions applied along the horizon of prediction. Next sections review ways to translate horizon-axis plots of accuracy into a single metric.

4.1 Predictor's Accuracy

First, we visit some metrics both from time series prediction and multi-label classification research streams, required if someone wants to establish comparisons with these works. Second, we introduce key metrics to cover the different accuracy perspectives for this problem.

Multistep-ahead Prediction

The simplest way of understanding the accuracy of a multistep-ahead predictor is to use the mean absolute error (MAE) or the simple mean squared error (MSE), the mean relative absolute error (RAR) or the average normalized mean squared error (NMSE), the ratio between the MSE and the time series variance. The normalized root mean squared error (NRMSE) either uses the series amplitude (when the attribute under prediction is numeric) or the number of labels (when the target attribute is an ordinal) to normalize the error. The accuracy is sometimes assessed through the symmetric mean absolute percentage of error (SMAPE) [8]. The average SMAPE over all time series under test is referred as SMAPE*. Other less frequent metrics, as the average minus log predictive density (mLPD) or relative root mean squared-error (RRMSE), are only desired for very specific types of datasets and, therefore, are not consider.

In fact, every similarity function can be used to compute a normalized distance error. A detailed survey of similarity-measures is done in [21]. Euclidean-distance, similarly to SMAPE, is simple and competitive. Dynamic Time Warping treats misalignments, which is important when dealing with long horizons of prediction. Longest Common Subsequence deals with gap constraints. Pattern-based functions consider shifting and scaling in both temporal and amplitude axis. These similarity functions have the advantage of smoothing error accumulation, but the clear drawback of the computed accuracy to not be easily comparable with literature results.

$$NMSE(y,\hat{y})=\frac{\frac{1}{h}\Sigma_{i=1}^{h}(y_i-\hat{y}_i)^2}{\frac{1}{h-1}\Sigma_{i=1}^{h}(y_i-\bar{y})^2}$$

$$NRMSE(y,\hat{y})=\frac{\sqrt{\frac{1}{h}\Sigma_{i=1}^{h}(y_i-\hat{y}_i)^2}}{y_{max}-y_{min}} \in [0,1]$$

$$SMAPE(y,\hat{y})=\frac{1}{h}\Sigma_{i=1}^{h}\frac{|y_i-\hat{y}_i|}{(y_i+\hat{y}_i)/2} \in [0,1]$$

To compute the predictor's accuracy, the multiple correlation coefficient R^2 is adopted. Both the average and the harmonic mean (minimizing the problems of the simple mean) are here proposed. A threshold for a set of testing instances below 0.9 is consider non-acceptable in many domains.

$$Acc_i(y,\hat{y})=1-(\text{NRMSE}(y,\hat{y}) \vee \text{SMAPE}(y,\hat{y}))$$

$$Accuracy=\frac{1}{n}\Sigma_{j=1}^{n}Acc_i(y^j,\hat{y}^j) \vee n(\Sigma_{j=1}^{n}\frac{1}{Acc_i(y^j,\hat{y}^j)})^{-1}$$

In sequence learning, additional accuracy metrics consider functions applied to subsequences. The simplest case is of boolean functions that verify the correct labeling of contiguous points. The variance of functions applied to subsets of contiguous periods is key if the performance of the predictor deteriorates heavily across the horizon of prediction. This metric, here referred as error accumulation, avoid the need of a visual comparison of accuracy across the horizon.

Multi-label Classification

Multi-label classification metrics are relevant to compare results when the class under multi-period prediction is nominal. Note that beyond the common intersection operator used to compute accuracy, additional functions are broadly adopted to differentiated costs for false positives and true negatives and to allow for XOR differences.

$Accuracy = \frac{1}{n}\Sigma_{j=1}^{n}\frac{|y^j \cap \hat{y}^j|}{|y^j \cup \hat{y}^j|}$ [53]

$HammingLoss = \frac{1}{n}\Sigma_{j=1}^{n}\frac{|y^j\ XOR\ \hat{y}^j|}{h}$ [53]

Target Accuracy Metrics

When the class for prediction is numeric or ordinal, the accuracy of the long-term predictor should follow one of the loss functions adopted in multistep-ahead prediction. Preferably, $NRMSE$ and $SMAPE$ if the goal is to compare with literature results. A similarity function that treats misalignments should be complementary applied for further understanding of the predictor's performance.

If the class for prediction is nominal, the accuracy should follow the adapted multi-label accuracy metric defined below, and be potentially complemented with other loss functions to deal with temporal labeling misalignments.

$Acc_i(y^j, \hat{y}^j) = \frac{1}{h}\Sigma_{i=1}^{h} \mid y_i^j \cap \hat{y}_i^j \mid \vee\ 1-\text{LossF}(y^j, \hat{y}^j)$

$Accuracy = \frac{1}{m}\Sigma_{j=1}^{m} Acc_i \vee n(\Sigma_{j=1}^{n}\frac{1}{Acc_i(y^j,\hat{y}^j)})^{-1}$

Accuracy may not suffice to evaluate long-term predictors. Specificity, sensitivity and precision can be evaluated recurring to a three-dimensional decision matrix, where, for instance, an harmonic mean can be applied to eliminate the temporal dimension.

In non-balanced datasets, as the target healthcare datasets, most of the considered instances are in a non-relevant category. For instance, critical patients are just a small subset of all instances. A system tuned to maximize accuracy can appear to perform well by simply deeming all instances non-relevant to all queries. In the given example a predictor that outputs zero hospitalizations for every patient may achieve a high accuracy rate. A deep understanding can be made by studying the sensitivity, fraction of correctly predicted instances that are relevant, and specificity, the fraction of relevant instances that are correctly predicted. F-measure trades-off precision versus recall in a single metric. By default, $\alpha = 1/2$, the balanced F-measure, equally weights precision and recall.

To redefine these metrics, a boolean criteria T is required to decide whether an instance is of interest. For example, relevant patients have average yearly hospitalizations above 2. Table 2 presents the confusion matrix for the target predictors, from which a complementary set of metrics were retrieved.

Table 2 Confusion matrix for long-term predictors

	Relevant	Non-relevant
Positive	tp=$\Sigma_{j=1}^{n} T(y) \wedge Acc(y,\hat{y}) \geq \beta$	fp=$\Sigma_{i=1}^{n}$(1-$T(y)) \wedge Acc(y,\hat{y}) \geq \beta$
Negative	fn=$\Sigma_{j=1}^{n} T(y) \wedge Acc(y,\hat{y}) < \beta$	tn=$\Sigma_{i=1}^{n}$(1-$T(y)) \wedge Acc(y,\hat{y}) < \beta$

$$Precision = \frac{tp}{tp+fp} = \frac{\Sigma_{j=1}^{n}(T(y^j) \wedge Acc(y^j,\hat{y}^j) \geq \beta)}{\Sigma_{j=1}^{n} Acc(y^j,\hat{y}^j) \geq \beta}$$

$$Recall = \frac{tp}{tp+tn} = \frac{\Sigma_{j=1}^{n}(T(y^j) \wedge Acc(y^j,\hat{y}^j) \geq \beta)}{\Sigma_{j=1}^{n} T(y^j)}$$

$$F_{Measure} = \frac{1}{\alpha\frac{1}{Precision} + (1-\alpha)\frac{1}{Recall}}, \text{ where } \alpha \in [0,1]$$

$$RoundAccuracy = \frac{tp+fp}{tp+tn+fp+fn} = \frac{1}{n}\Sigma_{j=1}^{n}(Acc(y^j,\hat{y}^j) \geq \beta)$$

4.2 Other Relevant Metrics

Predictor's *error accumulation*, the propagation of past prediction errors into future predictions, can be expressed by a bias-variance for squared loss functions as defined in [17].

Predictor *utility* defines the interestingness of long-term predictors based on usefulness, novelty and understandability metrics. Usefulness concerns the probability of an arbitrary instance to have their unlabeled multi-points classified according to a well-defined behavior. Novelty measures the contribution of a predictive model to increase the knowledge of the domain. Finally, understandability refers to the ability of retrieving knowledge from the learner. This work does not consider utility due to domain-driven multiplicity of usefulness and novelty criteria [1] and as consequence of the increasing availability of methods to achieve high understandability [43].

Finally, *smoothness* metrics [17] evaluate the ability of the predictor avoid over-fitting when noise fluctuations are present.

4.3 Data Properties

The repositories to test the target predictors may follow a multi-dimensional structure, a multivariate temporal structure, a tabular structure denormalized from a dimensional structure (i.e., implicitly contains temporality), or an entity-relationship structure (easily mapped into multi-dimensional models for a specific mining goal). The domain and properties of the adopted datasets (either multi-dimensional, relational or series-based) should be made available. Variables include degree of sparsity, noise sensitivity, completeness, length, degree of stationarity, presence of static features, discretization constraints, sensitivity to temporal shifts, and, in the case of temporal structures, allowance for itemsets, multivariate order and alphabet amplitude.

5 Related Research

Work on active research streams, illustrated in Fig.3, have presenting important results to constrain the solution space.

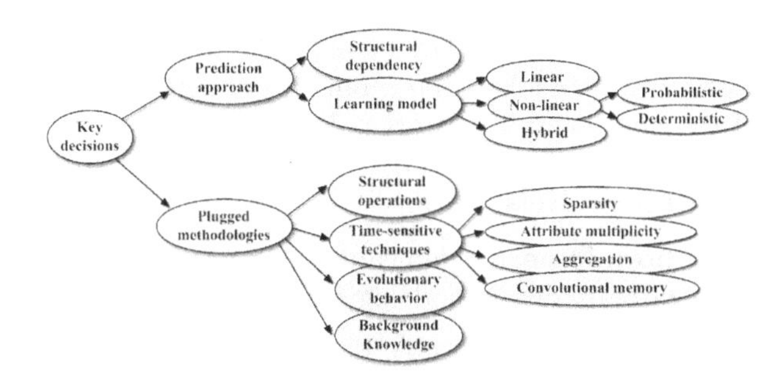

Fig. 3 Key areas for the definition of principles across different settings

5.1 Prediction Approach

A. Structural Dependency. A decision axis is whether to consider or not dependencies among the periods under prediction. Conventional approaches follow a multiple-input single-output mapping. In *iterated* methods [10][8], a h-step-ahead prediction problem is tackled by iterating, h times, the one-step-ahead predictor. Taking estimated values as inputs has an evident negative impact in error propagation [46]. *Direct* methods perform the h-step-ahead prediction by learning h models, each returning a direct forecast. Although not prone to error accumulation [46], they require higher functional complexity to model the stochastic dependencies between two non-similar series. Additionally, the fact that the n models are learned independently, prevents this approach from considering underlying dependencies among the predicted variables that may result in a biased learning [11]. In literature, successful *hybrids* that combine both approaches exist [47].

Multiple-Input Multiple-Output (MIMO) methods learn one model that preserves the stochastic dependencies for a reduced bias, although reduces the flexibility and variability of single-output approaches that may result in a new bias [11][8]. To avoid this, intermediate configurations can be imagined by decomposing the original task into $k = h/s$ tasks, each output with size s, where $s \in \{1, ..., n\}$. This approach, *Multiple-Input Several Multiple-Outputs* (MISMO), trades off the property of preserving the stochastic dependency among future values with a greater flexibility of the predictor [52].

B. Learning Model. Independently of the structural dependency choice, several learning paradigms exist. All of them, either implicitly or explicitly, model the multivariate conditional distribution $P(Y|A)$.

Learners can either follow linear or non-linear predictive models. Linear models include simple, logistic or Poisson regression, as integrated recurrent auto-regressions and feed-forward moving average mappings [34].

Non-linear long-term predictors can either define probabilistic or deterministic models. Most are adaptations of traditional classifiers using temporal sliding windows. Probabilistic predictors include (hidden) Markov models (HMM) [35], variable-memory Markov models [5], conditional random fields [30], and stochastic grammars [16]. Deterministic predictors include recurrent, time-delay and associate neural networks [25][31], multiple adaptive regression splines [36], regression and model trees [13][44], support vector machines (SVM) [15], and genetic solvers [18].

C. Plugged Methodologies. Significant performance improvements are triggered by plugged temporal methodologies that predictors may adopt [43].

C1. Structural Operations. Suitable dataset representations, similarity-measures and time-partitioning strategies are required for a quick and flexible learning. Criteria for temporal partitioning include clustering, user-defined granularities, fuzzy characterization, split-based sequential-trees, domain-driven ontologies and symbolic interleaving [42][40][1].

C2. Time-Sensitive Techniques. Strategies to enhance the performance of long-term predictors for healthcare planning tasks are required to answer the introduced requirements. Techniques to deal with *data sparsity* have been proposed, for instance, in [38][27]. The goal is to avoid the exponential growth of the target data structures and to correctly interpret empty time points. Time windows and feature-based descriptions have been proposed to deal with *temporal granularity*. [23] and [2] provide initial principles for hierarchical temporal zooming operations and calendars.

Techniques to deal with *data attributes multiplicity* have been addressed in multivariate response prediction research streams. The task is to predict a matrix of responses based for multivariate time series [14]. Due to the complexity of this task, existing solutions are linear vector auto-regressions

[32]. Although multivariate responses are useful to assist the prediction of class-of-interest, content-temporal dependencies among the input attributes are not considered.

Finally, covariance functions to deal with *memory sampling*, following either a parametric or non-parametric approach for the selective retaining of decisive events have been proposed in [50]. Strategies, as binary or exponentially decaying weighted of an input function, set a trade-off between *depth* (how far memory goes) and *resolution* (degree of data preservation).

C3. Evolutionary Behavior. The understanding of evolving behavior to balance the smoothing and overfitting problems of long-term predictors is still a youth research stream. Prediction rules, which specify a causal and temporal correlation between time points, have been used to assist the predictors learning [43]. In [22], emerging or evolutionary patterns, patterns whose support increases significantly over time, are adopted.

C4. Background Knowledge. Finally, background knowledge is increasingly claimed as a requirement for long-term prediction, as it guides the definition of time windows [42]; provides methods to bridge different time scales, to treat monitoring holes and to remove domain-specific noise [6]; defines criteria to prune the explosion of multiple-equivalent patterns [3]; and fosters the ability to incrementally improve results by refining the way domain-knowledge is represented [3]. A hierarchy of flexible content constraints, and of taxonomical and relational time relaxations is given in [1]. Further modeling of domain-driven temporal dynamics is required [2].

5.2 Related Research Streams

Time series long-term prediction, sequence learning and multi-label classification are the research streams with major relevant contributions.

Long-Term Prediction. Although traditionally applied over time series, it can be extended to deal with time sequences.

In [17], a comparative study on the performance of iterated, direct and hybrid single-output approaches in terms of their error accumulation, smoothness of prediction, and learning difficulty is done. Selected literature have been provided different methods to define s-variable in MISMO approaches (as cross-validation for different values or as a function of the current query point). Experimental studies [52] show that the choice of s strongly varies according to the case, with s=1 (Direct method) and s=n (MIMO method) being good performers in less than 20% of the cases. Improvements have been achieved in case of a large horizon h by adopting time series operators as the total or partial autocorrelation in multiple-output approaches [52]. A comparison of five multi-step-ahead prediction methods is done in [8].

Linear AIRMA models have been applied to deal with non-stationarity by defining a separated model learning for each suitable temporal window assumed to be stationary. Alternatively, in [45], clustering is combined with linear function approximation. In [17] a non-linear probabilistic predictor, an hybrid HMM-regression, is evaluated using different regression orders and predicting windows sizes. Regression trees [13] and model trees [44] are adapted decision trees, where each leaf stores a linear predictor.

Evaluation of three multiple-output neural network predictors (simple feed-forward, modular feed-forward and Elman) is done in [9]. In [39] temporal convolution machines use Gaussian distributions to learn a class of multimodel distributions over temporal data and are implemented using three recurrent neural network variants. In [12], Bayesian learning is applied to deal with noisy and non-stationary time series.

In [11], multiple-output approaches are extended with query-based criteria grounded on local learning. In [29], least-squares support-vector-machines (LS-SVM) are adopted with a local criteria for input selection, mutual information, to estimate dependencies according to Shanon entropy principle. In [46], k-nearest neighbors selection and noise estimation are additional criteria applied to select parameters to guide ARIMA and neural networks with encouraging results.

Sequence Learning. Sequence learning methods are adopted when the mining goal is sequence prediction, sequence recognition or sequential decision making [48]. The sequence recognition problem can be formulated as a prediction problem, $\hat{\mathbf{y}} = \mathbf{y}$ where $\hat{\mathbf{y}} = M(a_1, .., a_m)$. In the field of sequential decision making, sequence elements represent system states and the goal is to compose actions, Z, to reach a specific state $P(Z_{A \rightarrow Y}|AY)$ or to satisfy a goal $P(Z_{G=true}|A)$ [48]. Only contributions to the sequence prediction problem will be considered. Although sequence prediction only considers the causal ordering of elements, it provides important principles to consider in the solution space.

Unsupervised and reinforcement learning techniques from machine learning have been applied to sequence prediction, although still not scalable for large data volumes. We will briefly cover these contributions as they define important principles to solve the introduced problem. Additionally, learning techniques as expectation maximization, gradient descendant, policy iteration, hierarchical structuring or grammar training can be transversally applied to different implementations [30].

First, *unsupervised learning* are required in long-term prediction to avoid a biased learning towards smoothing or overfitting, and to deal with temporally non-compliant instances. Motifs, calendric rules, episodes, containers and partially-ordered tones [1][43][40] may be patterns of interest to assist prediction. Different approaches for their use within predictors exits. In [37], patterns are translated into boolean features to guide SVMs and logistic regressions.

Second, *reinforcement learning* [51] with two major types of predictors: *inductive logic predictors* that learn symbolic knowledge from sequences in the form of expressive rules [33], and *evolutionary computing predictors* that use heuristic-search over probabilistic models of pattern likelihood [41]. Both methods are applied with temporal-difference methods [51]. These techniques are the preference when one is not interested in a specific temporal horizon, but rather in predicting the occurrence of a certain symbol or pattern. In [20], sequence-generating rule models are defined to place constrains on which symbol can appear. In [19], time series are discretized into feature vectors to train trees by varying parameters as the width of the sliding window, from which rules are retrieved and combined with logical operators.

A large spectrum of implementations are, in fact, hybrid predictors. Examples include the use of symbolic rules and evolutionary computation applied to neural networks [49]. Although formal rule-based languages obtained by induction can be used for long-term prediction, these methods have not been extensively applied due to inference complexity [43].

Multi-label Classification. In [53] an overview of simple and hierarchical multi-label classifiers is done. Multi-label learning provide basic principles to deal with the long-term classification of nominal classes. Five methods that transform the multi-label classification problem either into single-label classification or regression problems are introduced. A set of classifiers and predictors are adapted for multi-label data. Examples include a revised C4.5 with an adapted entropy calculation [53], a kNN lazy learner that includes label-ranking probabilities [28], an extended AdaBoost and a novel probabilistic generative model [24].

6 Discussion

This work formalizes the problem of long-term prediction over multi-dimensional structures. It discusses the novelty and relevance of the problem in real-world applications. Accuracy, error propagation, noise sensitivity and complementary metrics to deal with non-balanced datasets were pointed as critical and defined.

Limitations and potential contributions are detailed from the three related research streams – multistep-ahead prediction, sequence learning and multi-label classification. Attribute multiplicity, conditional-dependency, and occurrence-sparsity are key challenges to solve the target problem. Empirical contributions, in the form of principles assessing one or more of these challenges, are the required next steps to promote an efficient learning of accurate predictors.

Acknowledgment. This work is supported by *Fundação para a Ciência e Tecnologia* under the project D2PM, PTDC/EIA-EIA/110074/2009, and the PhD grant SFRH/BD/75924/2011.

References

1. Antunes, C.: Pattern Mining over Nominal Event Sequences using Constraint Relaxations. Ph.D. thesis, Instituto Superior Tecnico (2005)
2. Antunes, C.: Temporal pattern mining using a time ontology. In: EPIA, pp. 23–34. Associação Portuguesa para a Inteligéncia Artificial (2007)
3. Antunes, C.: An ontology-based framework for mining patterns in the presence of background knowledge. In: ICAI, pp. 163–168. PTP, Beijing, China (2008)
4. Bacchus, F., Kabanza, F.: Using temporal logics to express search control knowledge for planning. A.I. 116, 123–191 (2000)
5. Begleiter, R., El-Yaniv, R., Yona, G.: On prediction using variable order markov models. J. Artif. Int. Res. 22, 385–421 (2004)
6. Bellazzi, R., Ferrazzi, F., Sacchi, L.: Predictive data mining in clinical medicine: a focus on selected methods and applications. Wiley Interdisc, DM and Knowledge Discovery 1(5), 416–430 (2011)
7. Bellazzi, R., Zupan, B.: Predictive data mining in clinical medicine: Current issues and guidelines. IJ Medical Information 77(2), 81–97 (2008)
8. Ben Taieb, S., Sorjamaa, A., Bontempi, G.: Multiple-output modeling for multi-step-ahead time series forecasting. Neurocomput. 73, 1950–1957 (2010)
9. Bengio, S., Fessant, F., Collobert, D.: Use of modular architectures for time series prediction. Neural Proc. Lett. 3, 101–106 (1996)
10. Bontempi, G., Birattari, M., Bersini, H.: Lazy learning for iterated time-series prediction. In: I.W. on A. Black-Box T. for Nonlinear Modeling, pp. 62–68. Katholieke U.L., Leuven (1998)
11. Bontempi, G., Ben Taieb, S.: Conditionally dependent strategies for multiple-step-ahead prediction in local learning. Int. J. of Forecasting 27(2004), 689–699 (2011)
12. Brahim-Belhouari, S., Bermak, A.: Gaussian process for nonstationary time series prediction. Computational Statistics and Data Analysis 47(4), 705–712 (2004)
13. Breiman, L., Friedman, J.H., Olshen, R.A., Stone, C.J.: Classification and Regression Trees. Chapman & Hall, New York (1984)
14. Brown, P.J., Vannucci, M., Fearn, T.: Multivariate bayesian variable selection and prediction. Journal of the Royal Statistical Society 60(3), 627–641 (1998)
15. Burges, C.J.C.: A tutorial on support vector machines for pattern recognition. Data Min. Knowl. Discov. 2, 121–167 (1998)
16. Carrasco, R.C., Oncina, J.: Learning Stochastic Regular Grammars by Means of a State Merging Method. In: Carrasco, R.C., Oncina, J. (eds.) ICGI 1994. LNCS, vol. 862, pp. 139–152. Springer, Heidelberg (1994)
17. Cheng, H., Tan, P.-N., Gao, J., Scripps, J.: Multistep-Ahead Time Series Prediction. In: Ng, W.-K., Kitsuregawa, M., Li, J., Chang, K. (eds.) PAKDD 2006. LNCS (LNAI), vol. 3918, pp. 765–774. Springer, Heidelberg (2006)
18. Cortez, P., Rocha, M., Neves, J.: A Meta-Genetic Algorithm for Time Series Forecasting. In: Proc. of AIFTSA 2001, EPIA 2001, Porto, Portugal, pp. 21–31 (2001)
19. Cotofrei, P., Stoffel, K.: First-Order Logic Based Formalism for Temporal Data Mining. In: Foundations of Data Mining and knowledge Discovery. SCI, vol. 6, pp. 185–210. Springer, Heidelberg (2005)
20. Dietterich, T.G., Michalski, R.S.: Discovering patterns in sequences of events. Artif. Intell. 25, 187–232 (1985)

21. Ding, H., Trajcevski, G., Scheuermann, P., Wang, X., Keogh, E.J.: Querying and mining of time series data: experimental comparison of representations and distance measures. In: Proceedings of the VLDB Endowment, vol. 1(2), pp. 1542–1552 (2008)
22. Dong, G., Li, J.: Efficient mining of emerging patterns: discovering trends and differences. In: 5th ACM SIGKDD, KDD, pp. 43–52. ACM, NY (1999)
23. Fang, Y., Koreisha, S.G.: Updating arma predictions for temporal aggregates. Journal of Forecasting 23(4), 275–296 (2004)
24. Freund, Y., Schapire, R.: A decision-theoretic generalization of on-line learning and an application to boosting. In: Proc. of the 2nd European Conf. on Comput. Learning Theory, pp. 23–37. Springer, London (1995)
25. Guimarães, G.: The Induction of Temporal Grammatical Rules from Multivariate Time Series. In: Oliveira, A.L. (ed.) ICGI 2000. LNCS (LNAI), vol. 1891, pp. 127–140. Springer, Heidelberg (2000)
26. Henriques, R., Antunes, C.: An integrated approach for healthcare planning over dimensional data using long-term prediction. In: 1st Proc. in Healthcare Inf. Systems. Springer, Beijing (2012)
27. Hsu, C.N., Chung, H.H., Huang, H.S.: Mining skewed and sparse transaction data for personalized shopping recommendation. Machine Learning 57, 35–59 (2004)
28. The IEEE Computational Intelligence Society: A k-Nearest Neighbor Based Algorithm for Multi-label Classification 2 (2005)
29. Ji, Y., Hao, J., Reyhani, N., Lendasse, A.: Direct and Recursive Prediction of Time Series Using Mutual Information Selection. In: Cabestany, J., Prieto, A.G., Sandoval, F. (eds.) IWANN 2005. LNCS, vol. 3512, pp. 1010–1017. Springer, Heidelberg (2005)
30. Kersting, K., De Raedt, L., Gutmann, B., Karwath, A., Landwehr, N.: Relational Sequence Learning. In: De Raedt, L., Frasconi, P., Kersting, K., Muggleton, S.H. (eds.) Probabilistic Inductive Logic Programming. LNCS (LNAI), vol. 4911, pp. 28–55. Springer, Heidelberg (2008)
31. Kleinfeld, D., Sompolinsky, H.: Associative neural network model for the generation of temporal patterns: Theory and application to central pattern generators. Biophysical J. 54(6), 1039–1051 (1988)
32. Koch, I., Naito, K.: Prediction of multivariate responses with a selected number of principal components. Comput. Statistical Data Analysis 54, 1791–1807 (2010)
33. Lavrac, N., Dzeroski, S.: Inductive Logic Programming: Techniques and Applications. Ellis Horwood, New York (1994)
34. Laxman, S., Sastry, P.S.: A survey of temporal data mining. Sadhana-academy. Proc. in Eng. Sciences 31, 173–198 (2006)
35. Laxman, S., Sastry, P.S., Unnikrishnan, K.P.: Discovering frequent episodes and learning hidden markov models: A formal connection. IEEE Trans. on Knowl. and Data Eng. 17, 1505–1517 (2005)
36. Lee, T.S., Chiu, C.C., Chou, Y.C., Lu, C.J.: Mining the customer credit using classification and regression tree and multivariate adaptive regression splines. Computational Statistics & Data Analysis 50(4), 1113–1130 (2006)
37. Lesh, N., Zaki, M.J., Ogihara, M.: Mining features for sequence classification. In: Proc. of the 5th ACM SIGKDD, pp. 342–346. ACM, NY (1999)
38. Liu, J., Yuan, L., Ye, J.: An efficient algorithm for a class of fused lasso problems. In: Proc. of the 16th ACM SIGKDD, KDD, pp. 323–332. ACM, NY (2010)

39. Lockett, A.J., Miikkulainen, R.: Temporal convolution machines for sequence learning. Tech. Rep. AI-09-04, University of Texas at Austin (2009)
40. Mannila, H., Toivonen, H., Inkeri Verkamo, A.: Discovery of frequent episodes in event sequences. IJ of DMKD 1, 259–289 (1997)
41. Meeden, L.A.: An incremental approach to developing intelligent neural network controllers for robots. IEEE Trans. on Sys. Man and Cyber. 26(3), 474–485 (1996)
42. Moerchen, F.: Tutorial cidm-t temporal pattern mining in symbolic time point and time interval data. In: CIDM. IEEE, Nashville (2009)
43. Mörchen, F.: Time series knowledge mining. W. in Dissertationen. G&W (2006)
44. Quinlan, J.R.: Learning with continuous Classes. In: 5th Australian Joint Conf. on Artificial Intelligence, pp. 343–348 (1992)
45. Sfetsos, A., Siriopoulos, C.: Time series forecasting with a hybrid clustering scheme and pattern recognition. IEEE Transactions on Systems, Man, and Cybernetics, Part A 34(3), 399–405 (2004)
46. Sorjamaa, A., Hao, J., Reyhani, N., Ji, Y., Lendasse, A.: Methodology for long-term prediction of time series. Neurocomput. 70, 2861–2869 (2007)
47. Sorjamaa, A., Lendasse, A.: Time series prediction using dirrec strategy. In: ESANN, pp. 143–148 (2006)
48. Sun, R., Giles, C.L.: Sequence learning: From recognition and prediction to sequential decision making. IEEE Intelligent Systems 16, 67–70 (2001)
49. Sun, R., Peterson, T.: Autonomous learning of sequential tasks: experiments and analyses. IEEE Transactions on Neural Networks 9(6), 1217–1234 (1998)
50. Sutton, R.S.: Learning to predict by the methods of temporal differences. Machine Learning 3, 9–44 (1988)
51. Sutton, R., Barto, A.: Reinforcement learning: an introduction. Adaptive Computation and Machine Learning. MIT Press (1998)
52. Taieb, S.B., Bontempi, G., Sorjamaa, A., Lendasse, A.: Long-term prediction of time series by combining direct and mimo strategies. In: Proc. of the 2009 IJCNN, pp. 1559–1566. IEEE Press, USA (2009)
53. Tsoumakas, G., Katakis, I.: Multi Label Classification: An Overview. IJ of Data W. and Mining 3(3), 1–13 (2007)

Mining Patterns from Large Star Schemas Based on Streaming Algorithms

Andreia Silva and Cláudia Antunes

Abstract. A growing challenge in data mining is the ability to deal with complex, voluminous and dynamic data. In many real world applications, complex data is not only organized in multiple database tables, but it is also continuously and endlessly arriving in the form of streams. Although there are some algorithms for mining multiple relations, as well as a lot more algorithms to mine data streams, very few combine the multi-relational case with the data streams case. In this paper we describe a new algorithm, *Star FP-Stream*, for finding frequent patterns in multi-relational data streams following a star schema. Experiments in the emphAdventureWorks data warehouse show that Star FP-Stream is accurate and performs better than the equivalent algorithm, FP-Streaming, for mining patterns in a single data stream.

1 Introduction

In many real world applications, complex data is not only organized in multiple database tables, but it is also continuously and endlessly arriving in the form of streams. For example, in a real-time sales analysis problem, we need to handle a fast stream of sales transactions, along with a much slower stream of customers (news and recurring ones), and a mostly static stream of products (for long periods of time). Consequently, processing and learning from multiple data streams have become a very important data mining task.

Although there are some algorithms for mining multiple relations, as well as a lot more algorithms to mine data streams, only few combine the multi-relational case with the data streams case [8]. Multi-relational data stream mining, MRDSM, is an emerging and inter-disciplinary area of data mining that encompasses both problems.

Andreia Silva · Cláudia Antunes
Department of Computer Science and Engineering
Instituto Supeior Técnico / Technical University of Lisbon
Lisbon, Portugal
e-mail: {andreia.silva,claudia.antunes}@ist.utl.pt

R. Lee (Ed.): Computer and Information Science 2012, SCI 429, pp. 139–150.
springerlink.com

Multi-relational data mining (MRDM)[4] is a fairly recent area that aims for learning from multiple tables, related somehow by foreign keys, in its original structure, i.e. without joining all the tables in one before mining. A commonly used structure for databases is a *star schema*, which is composed of a central fact table linking a set of dimension tables. In a star schema, data is modeled as a set of facts, each describing an event or occurrence, characterized by a particular combination of dimensions. In turn, each dimension aggregates a set of attributes for a same domain property or constraint [9]. In recent years, the most common mining techniques for a single table have been extended to the multi-relational context, but there are few dedicated to star schemas ([2, 12, 14, 13]).

Frequent pattern mining over data streams [10] is also a very important area of data mining that allows us to handle large amounts of data, which may arrive continuously and endlessly. Its main ideas are to avoid multiple scans of the entire datasets, optimize memory usage, and use a small constant time per record. Existing techniques are able to avoid large amounts of data at a time, keeping only the needed information in some summary data structure. Most of the existing algorithms for mining data streams are designed for a single data table ([11, 6, 10]).

In this paper we describe a MRDSM method, *Star FP-Stream*, for finding frequent patterns in multi-relational data streams modeled as a star schema (star streams). The algorithm is an extension of the data streams' algorithm *FP-Streaming* [6] based on *FP-Growth* [7], and adopts the "mining before join" strategy of the MRDM algorithm *Star FP-Growth* [13].

In section 2 we define the problem of mining frequent itemsets in star streams, and then we present the related work, in section 3. The proposed method is described in section 4. Experimental results are shown and discussed in section 5 and section 6 concludes the paper with a discussion about the results achieved and some guidelines for future work.

2 Problem Statement

Frequent pattern mining aims for enumerating all frequent patterns that conceptually represent relations among discrete entities (or *items*). Depending on the complexity of these relations, different types of patterns arise, with the transactional patterns being the most common. A *transactional pattern* is just a set of items that occur together frequently.

Let S be a tuple (D_1, D_2, ..., D_n, FT) representing a data warehouse modeled as a star schema, with D_i corresponding to each dimension table and FT to the fact table.

Also, let $I = \{i_1, i_2, \dots, i_m\}$ be a set of distinct literals, called *items*. In the context of a database, an *item* corresponds to a proposition of the form (*attribute*, *value*), and a subset of items is denoted as an *itemset*. $T = (tid, X)$ is a tuple where tid is a tuple-id (corresponding to a primary key) and X is an itemset in I. Each dimension table in S, is a set of these tuples. Transactions on the fact table, from now on called *facts*, are sets of n *tids*: tuples of the form $(tid_{D_1}, tid_{D_2}, \dots tid_{D_n})$.

The *support* (or occurrence frequency) of an itemset X, is the number of transactions containing X in the database S. X is frequent if its support is no less than a predefined minimum support threshold, σ. In a database modeled as a star schema, where there are several tables, we must distinguish between the support considering just a single table versus the support considering all the database:

The *local support* of an itemset X, with items belonging to a table D_i, is the number of occurrences of X in D_i (D_i.localSup(X)).

The *global support* (or just *support*) of an itemset X is the number of facts that contain the *tid*s of the dimensional tuples that contain X (given by $tid(X)$), as in equation 1:

$$globalSup(X) = \sum_{tid}^{tid(X)} FT.localSup(tid) \tag{1}$$

The problem of multi-relational frequent pattern mining over star schemas is to mine all itemsets whose global support is greater or equal than $\sigma \times |FT|$, where $|FT|$ is the number of facts and $\sigma \in]0,1]$ the user defined *minimum support threshold*.

The previous definitions consider that the database is mined all together. Let us now assume that the tables are data streams, where new transactions arrive sequentially in the form of continuous streams. In this *star stream*, only the fact stream connects the other tables / streams. Let the *fact stream* $FS = B_1 \cup B_2 \cup ...B_k$ be a sequence of batches, where B_k is the current batch, B_1 the oldest one, and each batch is a set of facts. Additionaly, let N be the current length of the stream, i.e. the number of facts seen so far. In this streaming case, we should only see one fact and one batch at a time, and only once. Therefore, we have to store information that allow us to keep track of all patterns, taking into account that current infrequent itemsets may become frequent later, as well as current frequent ones may no longer be, after mining the following batches. As it is unrealistic to hold all streaming data in the limited main memory, and if we assume a *deterministic* approach (with no probability of failure), data streaming algorithms have to sacrifice the correctness of their results by allowing some counting errors. These errors are bounded by a user defined *maximum error threshold*, $\varepsilon \in [0,1]$, such that $\varepsilon \ll \sigma$. Thus, the support calculated for each item is an approximate value, which at most has an error of εN.

The problem of multi-relational frequent pattern mining over star streams consists in finding all itemsets whose estimated support is greater or equal to $(\sigma - \varepsilon) \times N$.

3 Related Work

There are many stand-alone algorithms to mine different types of patterns in traditional databases (with no streaming data and only one table), with *FP-growth* [7] one of the most efficient. This algorithm follows a pattern-growth philosophy that adopts a divide and conquer approach to decompose both the mining tasks and the databases. The algorithm represents the data into a compact tree structure, called *FP-tree*, to facilitate counting the support of each set of items and to avoid expensive, repeated database scans. It then uses a *depth-first search* approach to traverse the tree and find the patterns.

Some of the traditional algorithms have been extended to be able to mine several tables, as well as to deal with data streams. In this work we will focus on the task of frequent pattern mining (PM).

3.1 Multi-relational PM over Stars

The work on multi-relational pattern mining over star schemas has been increasing in the last years. In order to deal with multiple tables, pattern mining has to join somehow the different tables, creating the tuples to be mined. An option that allows the use of the existing single-table algorithms, is to join all the tables in one before mining (a step also known as propositionalization or denormalization). However, there are several disadvantages, like the possible explosion of attributes and null values, and the difficulty of this task when dealing with complex relations between the tables. Research in this area has shown that methods that follow the philosophy of *mining before joining* outperforms the methods following the *joining before mining* approach, even when the latter adopts the known fastest single-table mining algorithms [12].

The first multi-relational methods have been developed by the *Inductive Logic Programming* (ILP) community about ten years ago (WARMR [3]) , but they are usually not scalable with respect to the number of relations and attributes in the database. Therefore they are inefficient for databases with large schemas. Another drawback of ILP approaches is that they need all data in the form of prolog tables.

Few approaches were designed for frequent pattern mining over star schemas: an apriori-based [1] algorithm is introduced in [2], wich first generates frequent tuples in each single table, and then looks for frequent tuples whose items belong to different tables via a multi-dimensional count array; [12] proposed an efficient algorithm that mines first each table separately, and then two tables at a time to find patterns from multiple tables; [14] presented *MultiClose*, that first converts all dimension tables to a vertical data format, and then mines each of them locally, with a closed algorithm. The patterns are stored in two-level hash trees, which are then traversed in pairs to find multi-table patterns; StarFP-Growth, proposed in [13], is a pattern-growth method, based on FP-Growth [7], that avoids the candidate generation processing. Its main idea is to construct an FP-Tree for each dimension (*DimFP-Tree*), accordingly to the global support of its items. These trees are compact representations of the patterns of the respective dimension. Then, the algorithm builds a Super FP-Tree, combining the FP-Trees of each dimension, accordingly to the facts, and at last calls the original FP-Growth with this tree to find multi-relational patterns.

3.2 Data Streams

In many real world applications, data appears in the form of continuous data streams, as opposed to traditional static datasets. A data stream is an ordered sequence of instances that are constantly being generated and collected. The nature of this streaming data makes the mining process different from traditional data mining in several aspects: (1) each element should be examined at most once and as fast as possible; (2) memory usage should be limited, even though new data elements are continuously arriving; (3) the results generated should be always available and updated; (4) frequency errors on results should be as small as possible. This implies the creation and maintenance of a memory-resident summary data structure (also called *synopsis data structure*), that stores only the information that is strictly necessary to avoid loosing patterns [10]. Hence, data stream mining algorithms have to sacrifice the correctness of their results by allowing some counting errors. Existing approaches are deterministic or probabilistic: an algorithm is *deterministic* if it only allows an error in the frequency counts, and is *probabilistic* if it also allows a probability of failure. In this work we will focus on deterministic algorithms, so that we can have the guarantee that all existing patterns are found and returned.

The first algorithm for mining all frequent itemsets over all streaming data was *Lossy Counting* [11]. It divides the data stream into batches with width $\lceil 1/\varepsilon \rceil$, so that the batch id exactly refers to the threshold εN (the maximum error). Each itemset is associated with its estimated frequency and its maximum error. At each batch boundary, all itemsets which frequency plus its maximum error is less than the current batch are discarded. In the end, itemsets with frequency of at least $(\sigma - \varepsilon)N$ are returned, and all true patterns are guaranteed to be reported.

Reference [6] presented a novel algorithm, *FP-Streaming*, that adapts FP-Growth [7] to mine frequent itemsets in time sensitive data streams. The stream is divided into batches and the arriving transactions are stored in a new FP-tree structure. At each batch boundary, the frequent patterns are extracted from that FP-tree by means of the FP-Growth, and the pattern-tree structure (called *FP-stream*) is updated and pruned. Each node in this tree represents a pattern (from the root to the node) and its frequency is stored in the node, in the form of a *tilted-time window table*, which keeps frequencies for several time intervals. The tilted-time windows give a logarithmic overview on the frequency history of each pattern, allowing the algorithm to address queries requesting frequent itemsets over arbitrary time intervals, rather than only over the entire stream (called a *landmark model*). FP-Streaming returns all patterns with an estimated frequency larger than $(\sigma - \varepsilon)N$, and guarantees, like Lossy Counting, that all frequent itemsets in N are delivered.

Some other algorithms were proposed to mine frequent itemsets in data streams (see [10] for a more exhaustive survey), but most of them are adaptations of the strategies applied in the algorithms above.

3.3 MRDM over Data Streams

To the best of our knowledge, there are only two works on multi-relational frequent pattern mining over data streams, both based on ILP.

In [5] *SWARM*, a Sliding Window Algorithm for Relational Pattern Mining over data streams, is proposed. SWARM is a deterministic approach for data streams, that iteratively generates and evaluates the candidates for each slide. It keeps the patterns in a *Set Enumerated tree* that enumerates all possible patterns, and stores a sliding vector in each node with the support of the respective patterns for each slide. New patterns lead to the expansion or update of the tree. When a new slide flows, the support vector is shifted to remove the expired support and the tree is pruned to eliminate unknown patterns. For dealing with multi-relational databases it is based on WARMR [3], that needs all data in prolog form, composed with predicates of variables and constants (atoms), each representing the relations in the database.

Finally, [8] presented *RFPS* (Relational Frequent Patterns in Streams), a probabilistic approach based on period sampling, for finding relational patterns over a sliding time window of a relational data stream. Since it is also based on WARMR [3], it needs the database in prolog form. RFPS is an apriori-based algorithm[1] that first generates and tests candidates with the help of a *Patterns Joint Tree* (with the possible refinements of atoms), and then maintains frequent patterns in a *virtual stream tree*, based on a periodical sampling probability.

4 Mining Star Streams

Star FP-Stream is a MRDM algorithm for mining multiple relational data streams. It is able to find approximate frequent relational patterns in large databases following a star schema, assuming that patterns are measured from the start of the stream up to the current moment (landmark model). *Star FP-Stream* combines the strategies of two algorithms: *Star FP-Growth* [13] (MRDM algorithm over star schemas) and FP-Streaming [6] (data streaming algorithm), both based on the traditional algorithm FP-Growth [7].

As referred in [9], dimensions are, by definition, smaller than the fact table. Therefore, we assume that all dimension tables are kept in memory, and only the fact table (the *fact stream*) is arriving in batches.

The fact stream is conceptually divided into k batches of $\lceil 1/\varepsilon \rceil$ transactions each, so that the batch id $(1..k)$ exactly refers to the threshold εN (the maximum error allowed is k, one per batch).

All items that appear more than once in a batch are frequent with respect to that batch, and potentially frequent (or *subfrequent*) with respect to the entire stream. Items that appear just once in a batch can be discarded because they are *infrequent*, and even if they reappear later in other batches and become frequent, the loss of support will not affect significantly the calculated support (error is less than k).

To mine the star, the algorithm uses *Star FP-Growth* [13] techniques. The first batch is processed separately with Star FP-Growth, so that we can fix the order of frequent items for all next batches. For each new batch, the idea is to build the DimFP-Trees for each dimension as new facts arrive, with the respective occurring transactions (local mining step). When a batch is completed, the DimFP-Trees are combined to form a Super FP-Tree (global mining step), which will contain the itemsets of all dimension that co-occur in the current batch. To combine the DimFP-Trees, we have to scan the facts of the batch a second time, otherwise we would not know which paths of the different DimFP-Trees occur together. However, a fact is just a set of *tids*, therefore this extra scan is not significant.

These DimFP-Trees can then discarded, and the Super FP-Tree can be mined to extract frequent itemsets. These patterns are maintained and updated in a *pattern-tree*, which is a data structure based on the FP-tree [7] that maintains crucial, quantitative information only about patterns, instead of itemsets. Since there are often a lot of sharing of frequent items among patterns, the size of the tree is usually much smaller than having them in a list or a table, and therefore, searching for an itemset in it is usually much faster.

In Star FP-Stream, each node in the pattern-tree stores a pattern (corresponding to the path from the node to the root), its estimated frequency and its maximum error (which is the batch id of its insertion in the tree, since it corresponds to the number of times it could have been discarded). This information allows the algorithm to stop mining super sets of infrequent itemsets and to prune of infrequent items from the tree.

The Super FP-Tree is mined using FP-Growth algorithm, with 3 pruning strategies: for each mined itemset, if it is in the pattern tree but its estimated frequency plus its maximum error is less than or equal to the current batch id, it is infrequent in N and it will be removed in the final pruning step, thus stop mining supersets (*type II pruning*). If it is not in the pattern tree and appears only once, do not add it to the pattern tree and stop mining its supersets (*type I pruning*), but if it appears more than once, it is frequent in this batch, therefore insert it in the pattern tree and continue mining. After mining the batch, all itemsets in the pattern-tree which estimated frequency plus its maximum error is less than or equal to the current batch id are discarded (*tail pruning*).

At any point in time, Star FP-Stream can be asked to produce a list of the frequent itemsets along with their estimated frequencies, by traversing the pattern-tree and delivering all itemsets which estimated frequency plus its maximum error is not less than σN.

4.1 Strengths and Weaknesses

As a data streaming algorithm, Star FP-Stream gives the following guarantees (like [11, 6]): All item(set)s whose true frequency exceeds σN are returned. There are no *false negatives*; No item(set) whose true frequency is less than $(\sigma - \varepsilon)N$ is returned; And estimated frequencies are *less* than true frequencies by at most εN. As a multi-relational algorithm for star schemas, Star FP-Stream mines the star directly, without materializing the join of the tables, and all multi-relational patterns are returned.

Like any algorithm, Star FP-Stream also has some limitations: It has to scan the facts twice, first to know which transactions of dimensions occur, and second to combine them in the end of a batch. However, a fact is just a set of *tids*, therefore the time needed for each scan and the memory needed to keep it, are not significant; The pattern-tree tends to be very large, since it has to keep all frequent and subfrequent patterns. Nevertheless, its size tends to be stable as the batches increase, and it is able to return the patterns for every minimum support $\sigma \gg \varepsilon$, anytime.

5 Performance Evaluation

This section presents the experiments conducted to evaluate the performance of our data streaming algorithm. Our goal is to evaluate the accuracy, time and memory usage, and to show that: (1) Star FP-Stream has a good accuracy and does not miss any real pattern; (2) mining directly the star is better than joining before mining.

We assume that we are facing a landmark model, where all patterns are equally relevant, regardless of when they appear in the data. Therefore, we test Star FP-Stream over an adaptation of FP-Streaming for landmark models, which we will call Simple FP-Stream, that stores only one counter in each node of the pattern tree (instead of one per time window). Since Simple FP-Stream does not deal with stars directly, it denormalizes each fact when it arrives (i.e. it goes to every dimension and join all the transactions corresponding to the *tid*s of the fact in question), before mining it.

Star FP-Growth was also implemented so that we can run it on all data and compare the returned patterns (the exact patterns) and evaluate the accuracy of Star FP-Stream results (approximate patterns).

Experiments were conducted varying both minimum support and maximum error thresholds ($\sigma \in \{50\%, 40\%, 30\%, 20\%, 10\%\}$ and $\varepsilon \in \{10\%, 5\%, 4\%, 3\%, 2\%, 1\%, 0.5\%\}$, respectively). A common way to define the error is $\varepsilon = 0.1\sigma$ [10]. (Additionally, we use a larger error to see how worse the results are, and a smaller error to see the improvements).

Note that the course of the mining process of streaming algorithms does not depend on the minimum support defined, only on the maximum error allowed. The support only influences the pattern extraction from the pattern-tree, which, in turn, is ready for the extraction of patterns that surpass any asked support ($\sigma \gg \varepsilon$).

We tested the algorithms with a sample of the *AdventureWorks 2008 Data Warehouse* (DW)[1], a DW created by *Microsoft*, especially to support data mining scenarios.

In this work we will analyze a sample of the star *Internet sales*, which contains information about more than 60 thousand individual customer Internet sales orders, from July 2001 to July 2004. Dimension tables were kept in memory and the fact table is read as new facts are needed. We will consider four dimension tables: *Product*, *Date*, *Customer* and *SalesTerritory*, so that we are able to relate who bought what, when and where. The fact table has only the keys of those four dimensions (other attributes were removed), and each dimension has only one primary key and other attributes (no foreign keys). Numerical attributes were excluded (except the year and semester in dimension Date) as well as translations, and other personal textual attributes, like addresses, phone numbers, emails, names and descriptions.

The computer used to run the experiments was an Intel Xeon E5310 1.60GHz (Quad Core), with 2GB of RAM. The operating system used was GNU/Linux amd64 and the algorithms were implemented using the Java Programming language (Java Virtual Machine version 1.6.0_24).

[1] Available at http://sqlserversamples.codeplex.com/

5.1 *Experimental Results*

The accuracy of the results is influenced by both error and support thresholds. The resulting patterns of Star FP-Stream and Simple FP-Stream are the same (the algorithms only differ in how they manipulate the data).

We know that as the minimum support decreases, the number of patterns increase, since we require fewer occurrences of an item for it to be frequent. And as the maximum error increases, the number of patterns returned also tends to increase, because although we can discard more items, we have to return more possible patterns to make sure we do not miss any real one. The *precision* measures the rate of real patterns over the patterns returned by the streaming algorithm.

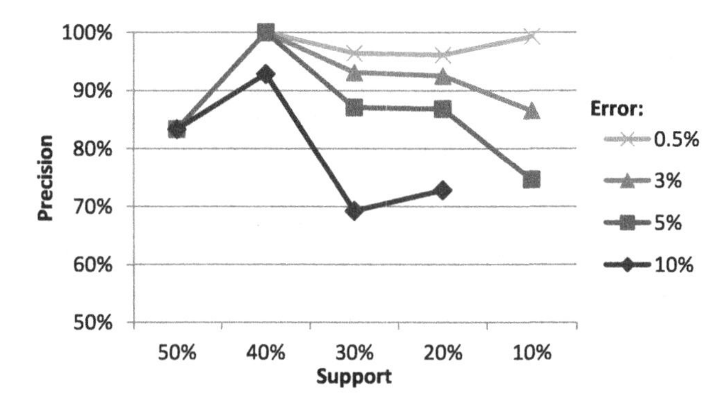

Fig. 1 Precision variation per support

Fig. 1 presents the precision as the support varies. These results depend on the data characteristics, namely in the number of hidden patterns and in the history of occurrences of items across the batches processed. In this case of *AdventureWorks*, we can state that for 40% of support the algorithm achieved better results (100% of precision for errors between 1% and 5%). This may mean that patterns that appear more than 40% of the times are well defined and consequently are monitored early during processing. We can also see that, as the error increases, the precision decreases, for all support thresholds. In other words, the smaller the error, fewer non real patterns are returned. The overall results show that precision is always above 69%.

The *recall* of Star FP-Stream (and Simple FP-Stream) is proved theoretically to be 100% (there are no false positives, i.e. there are no real patterns that the algorithm considers infrequent).

5.1.1 Time

Processing time was analyzed in terms of the time needed to process one batch (update time). It consists on the elapsed time from the reading of a transaction to the update of the pattern tree.

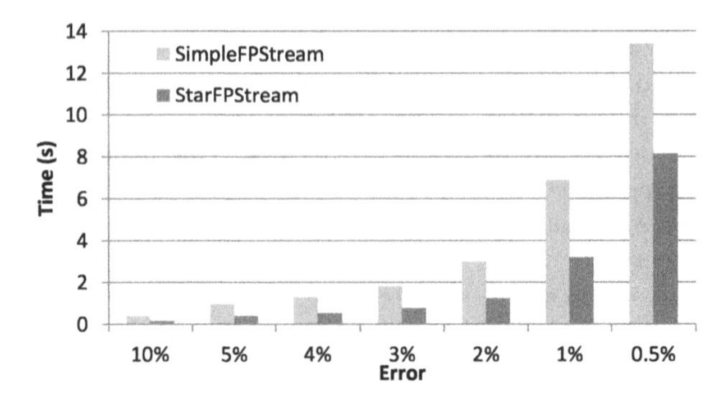

Fig. 2 Average update time

Fig. 2 shows the average update time of both algorithms for all errors. For consistency, we do not take into account the time needed to process the first batch, since it is processed separately.

We can state that Simple FP-Stream demands, on average, more time than Star FP-Stream. This demonstrates that, for star streams, denormalize before mining takes more time than mining directly the star schema, corroborating our goal and one of the goals of MRDM.

The update time is required to be constant and not depend on the number of transactions. By analyzing in detail the time needed per batch, we verified that the update time tends to be constant and does not increase as more batches are processed. In the first batches, there are a lot of patterns in the pattern tree, and both algorithms need more time to look for and to add patterns to the tree. Around the 5500 transactions, the data allows the algorithms to prune almost half of the patterns stored in the pattern tree, and then it keeps constant thereafter.

5.1.2 Memory

The space or memory used by the algorithms was also studied. It depends on the intermediate structures used by the algorithms, and it is strongly related with the size of the pattern tree (and therefore with the error bound). To analyze the maximum memory per batch, we measured the memory used by the algorithms for each batch, right before discarding the Super FP-Tree and doing the pruning step. We noted that the algorithms perform very similar. Star FP-Stream needs a bit more memory per batch than Simple FP-Stream, which was expected, since the first has to construct the DimFP-trees for each dimension, while the second puts the denormalized facts in just one FP-tree. Although Star FP-Stream needs, in average, a bit more memory per batch than Simple FP-Stream, it needs less memory in the worst cases. Just as in the time usage, the memory needed is higher in the first batches, and then it tends to stabilize. In the first batches, Star FP-Stream performs better in terms of memory than Simple FP-Stream, but later on, it tends to need a little more memory, for most errors. The explanation for the higher values in the first batches is the same as for the time needed.

6 Conclusions and Future Work

In this paper, we propose a new algorithm for mining very large data warehouses, modeled as star schema, by combining Star FP-Growth and FP-Streaming.

Experiments on Adventure Works allowed us to analyze the behavior of our algorithm, comparing its performance for several errors and supports in terms of its accuracy, time and memory needed. Results indicate that Star FP-Stream is accurate, achieving a good precision and 100% of recall. The time and memory needed by the algorithms tend to be constant and do not depend on the total number of transactions processed so far, but only on the size of the batches and on the size of the current pattern tree, which in turn depends on the characteristics of the data.

Star FP-Stream is the algorithm that performs better. Although it needs a bit more memory per batch, it is not substantial, and it needs significantly less time to process each batch, thereby overcoming the "join before mining" approach.

From this point, there are two issues to solve. First, it is important to be able to deal with measures recorded in fact tables, for example by propagating their values for influencing support. The second issue relates to degenerated dimensions, which are just used for aggregating facts. In this case, it is necessary to redefine the notion of support, since it may be different according to distinct aggregations.

Acknowledgment. This work is partially supported by FCT – Fundação para a Ciência e a Tecnologia, under research project educare (PTDC/EIA-EIA/110058/2009) and PhD grant SFRH/BD/64108/2009.

References

1. Agrawal, R., Srikant, R.: Fast algorithms for mining association rules in large databases. In: VLDB 1994: Proc. of the 20th Intern. Conf. on Very Large Data Bases, pp. 487–499. Morgan Kaufmann, San Francisco (1994)
2. Crestana-Jensen, V., Soparkar, N.: Frequent itemset counting across multiple tables. In: PADKK 2000: Proc. of the 4th Pacific-Asia Conf. on Knowl. Discovery and Data Mining, London, pp. 49–61 (2000)
3. Dehaspe, L., Raedt, L.D.: Mining Association Rules in Multiple Relations. In: Džeroski, S., Lavrač, N. (eds.) ILP 1997. LNCS, vol. 1297, pp. 125–132. Springer, Heidelberg (1997)
4. Džeroski, S.: Multi-relational data mining: an introduction. SIGKDD Explor. Newsl. 5(1), 1–16 (2003)
5. Fumarola, F., Ciampi, A., Appice, A., Malerba, D.: A sliding window algorithm for relational frequent patterns mining from data streams. In: Proc. of the 12th Intern. Conf. on Discovery Science, pp. 385–392. Springer (2009)
6. Giannella, C., Han, J., Pei, J., Yan, X., Yu, P.S.: Mining frequent patterns in data streams at multiple time granularities: Next generation data mining (2003)

7. Han, J., Pei, J., Yin, Y.: Mining frequent patterns without candidate generation. In: SIGMOD 2000: Proc. of the 2000 ACM SIGMOD, pp. 1–12. ACM, New York (2000)
8. Hou, W., Yang, B., Xie, Y., Wu, C.: Mining multi-relational frequent patterns in data streams. In: BIFE 2009: Proc. of the Second Intern. Conf. on Business Intelligence and Financial Engineering, pp. 205–209 (2009)
9. Kimball, R., Ross, M.: The Data warehouse Toolkit - the complete guide to dimensional modeling, 2nd edn. John Wiley & Sons, Inc., New York (2002)
10. Liu, H., Lin, Y., Han, J.: Methods for mining frequent items in data streams: an overview. Knowl. Inf. Syst. 26, 1–30 (2011)
11. Manku, G.S., Motwani, R.: Approximate frequency counts over data streams. In: VLDB 2002: Proc. of the 28th Intern. Conf. on Very Large Data Bases, pp. 346–357. Morgan Kaufman, Hong Kong (2002)
12. Ng, E.K.K., Fu, A.W.C., Wang, K.: Mining association rules from stars. In: ICDM 2002: Proc. of the 2002 IEEE Intern. Conf. on Data Mining, pp. 322–329. IEEE, Japan (2002)
13. Silva, A., Antunes, C.: Pattern Mining on Stars with FP-Growth. In: Torra, V., Narukawa, Y., Daumas, M. (eds.) MDAI 2010. LNCS, vol. 6408, pp. 175–186. Springer, Heidelberg (2010)
14. Xu, L.J., Xie, K.L.: A novel algorithm for frequent itemset mining in data warehouses. Journal of Zhejiang University - Science A 7(2), 216–224 (2006)

Frameworks for the Effective Information Exchange on Mobile Devices

Haeng-Kon Kim and Roger Y. Lee

Abstract. Recently fast innovation of Internet technology causes lot of application to change into mobile application and the technology trends of communication equipment are changed from mono-function to multi-functioned system. These trends are part of changes which is caused by ubiquitous world and it is just beginning of huge waves which is required to fit and change under the ubiquitous environments. In this paper, we focused on the Design and Implementation of Component Objects that can be communicated effectively among various types of clients under the Heterogeneous Client Server Environments and Material Management System was chosen as the target of application. The key point to do that kind of affair is using component objects for the enforcement of reusability and inter-operability among and using XML mobile services that can communicate thru systems. Thus the Components proposed in this paper could be reused effectively in case of developing similar applications.

Keywords: component, object, layers, XML, client, server, mobile service, service.

1 Introduction

Recent years, most of the enterprises are forced to support various information to the their customers and those enterprises must support the information not only using their own developed software but also other information developed by third parties. In these cases software and data to interface with other systems could be increased

---

Haeng-Kon Kim
School of Information and Technology, Catholic University of Daegu, 712-702, Korea
e-mail: hangkon@cu.ac.kr

Roger Y. Lee
Software Engineering Information Technology Institute, Central Michigan University, USA
e-mail: leelry@cmich.edu

R. Lee (Ed.): Computer and Information Science 2012, SCI 429, pp. 151–163.
springerlink.com

enormously and the methodology to communicate and sharing information with other type of system is shown its head as important issue. These kind of problems cannot be solved by the traditional development method. Conventional procedure-oriented methodology that the enterprise applied causes lot of money comparing with the estimated costs[1]. For the more, even though the development had been finished, it requires much costs and time for the operation and maintenance of the system. Those kind of problems should be reduced efficiently in the ubiquitous ages because most of enterprises normally develop the products by the cooperation of other groups not only by themselves and the interoperability of the products, information might be more important for the growing various type of applications[2][9]. Instead of traditional methodology, CBD could be a good counter-proposal because it strengthen the reusability and expandability. However mobile service function is considered as the important methodology of internet application while business environments become more complicated and organized various type of heterogeneous systems. Thus more compact and component oriented software should be developed and applied to the business. This thesis is focused on the design and implementation of the component objects which can communicate with other clients under the heterogeneous environment and we have chosen Material Information System as target application area.

2 Related Works

The development process of software architecture covers a set of activities whose nature and ordering depend on the particular system, on the designers' skills, and on the tools available. Some of these activities are performed by hand while dedicated tools support others. In any case, the key activities in software architectural design include: functional and modular decompositions, functions allocation to modules, processes identification, mapping modules with processes and mapping processes with processors. These activities bring to bear architectural design guides such as software architecture reference models, architecture styles and patterns, and result in the production of a conceptual architecture followed by an implementation architecture. Figure 1 shows the design process for software architecture ranging from reference models to implementation architecture.

2.1 Layered Architecture

CBD concept was set up during 1990s decade and generally use layered architecture in building business software construction. Basic architecture of client/server was spread over the world in the beginning of 1990s and the concept of 3- layered architecture like User Interface layer, Business Logic Layer, Database Layer became the main trends after two layered architecture in the business industry[3][8]. Recently Business Logic Layer is apt to divided into Business Process Layer and Data Access Layer. However the Business Logic Layer is much complicated comparing with other layers, thus, lot of developers are concerning to that layer rather than others [4].

2.2 Middleware

After the end of 1990s, development of CBD is strengthen by the middleware above all. COM+ of Microsoft and J2EE of SUN that are considered as main stream in the business industry suggested MTS and EJB container as middleware. These middleware support the requirements of Business Component processed in server side. For example DB connectivity, calling among components, TP monitoring, load balancing and so on. Especially COM+ support interoperability among components without concerning to the language and J2EE support the implementation of components in heterogeneous operating system [5].

2.3 Mobile Service

Recently technology of Mobile Service is upgraded rapidly and it is forecasted by experts to affect revolutionary changes in the business industry. Mobile Service could be defined Loosely coupled software component technology of which service is supported by International standard technology like XML, SOAP and internet[6]. The outcome of Mobile Service make easy construction of information system and strengthen merits of CBD. First of all, Mobile Service support complete interoperability regardless platform type that is executing components and this approach is much better than COM+ or EJB. That is because of the standard internet technology like HTTP, SOAP, XML[7].

2.4 Modelling for Technical Design

The three approaches approaches presented above all provide methodological guidance for turning the results of the modeling activities into a technical design. In this sense, they build the aspects that have been modeled into the system. The process modeling approach takes its point of departure in the way users work. This relates more generally to a focus on this domain:

- *Application domain* : The individual persons or roles and the organization that administrates, monitors, or controls a problem domain

The application domain is where the users are and do whatever they do when they use the system. For an air traffic control system, the application domain is in the control tower where the controllers perform their air traffic control. The controllers monitor the traffic on the screen, decide on interventions, and direct the flights in their air space. With the process modeling approach, the domain-dependent aspects are elicited from the application domain and built into the system through the activities in which the software functions are designed. The data modelling approach takes a different point of departure by focusing on the data that people work with in the user organization. It has been argued that this was a much more stable foundation for software design than the way in which the users worked [8]. The data modeling approach relates more generally to a focus on this domain:

- *Problem domain* : The part of the context that is administrated, monitored or controlled by a system.

The problem domain is part of what is outside the system (i.e., in the context). For an air traffic control system the problem domain is that part of the context constituted by flights, departures, aircrafts, aircrafts positions and trajectories, changed altitude, changed speed, etc. Everything that the controller in the tower needs to know about to control the air space effectively is in the problem domain. It is fundamental to air traffic control that the controller watches the aircrafts positions and trajectories on a large monitor displaying data from the radar system rather than looking out the towers windows with a pair of binoculars. The system creates and maintains the controllers view of their air space and it is there crucial that the model of the air space (i.e., the problem domain) is in accordance with the controllers professional language and competence. With the data modeling approach, the domain-dependent aspects are elicited from the problem domain and built into the system through the activities in which the database and the related software are designed. The object modeling approach is more varied. Some of the methods, in particular the early ones, are focusing on the problem domain[9]. The RUP methods is completely opposite as it departs from use cases which are descriptions of the application domain. Rumbaugh[12] is the only classic object-oriented method that emphasizes both the problem domain and the application domain. Two of the three fundamental models are the class diagram, emphasizing the problem domain, and a description of functions by means of data-flow diagrams from structured analysis, emphasizing the application domain. This dual focus is an interesting and innovative approach. Unfortunately, the description of functions is not related to the object-oriented model. System developers with experience using the Rumbaugh method also point out that constructing the functional model is rarely worth while.

2.5 Software for Mobile Systems

The overview above illustrate that popular software engineering methods have a strong focus on technical aspects and the representation of information in the system. Yet they have very little in particular to offer in modelling context for mobile systems. Rational Unified Process[10], for example, offers several principles of which none address how to model the context of a mobile system. Microsoft Solutions Framework[11], as another example, offers a set of principles for software engineering, but, again, has nothing in particular to say on modeling the context of a mobile system. The literature on human-computer interaction has a stronger emphasis on the context of computerized systems. The basic literature deals with user interface design from a general point of view. They provide extensive guidelines and techniques for user interaction design but nothing specific on design of mobile systems and very little on modelling of domain-dependent aspect as a basis for technical design. Some of the literature in human-computer interaction deals specifically with user interaction design for mobile systems. There is a general textbook on design of user interaction for mobile systems. There is some literature that deals with technical design of context-aware systems. Lei and Zhang [12] apply the theory of conceptual graphs to mobile context modelling. All items in the context are monitored and the conceptual graphs model this as simple graphs associated with rules and constraints. Baumeister et al. extend UML with mobile objects, locations, and mobile activity. They formulate the extension in terms of stereotypes and end by providing a modified metamodel for UML.

3 Design of Effective Information Exchange on Mobile Devices

3.1 Basic Structure

The basic structures of this design are divided into 3 tiers. Those are Presentation Tier, Business Tier, and Data Tier. The Presentation Tier is divided into two type of components; UI component which handle user interface like I/O forms and UI Process Component which support the processing of UI Component. Business Tier is also divided into two type of component; Service Interface which transfer the request received from UI Process Component to the Business Entity, and the Business Entity perform the function of accessing Data Tier. All of the objects and output from Business Tier are transformed into XML typed structure. Thus it can communicate with heterogeneous type of clients. Among those tiers Business Tier should be examined carefully. Other tiers do not care the type of platform and the structure but Business Tier itself cant be like others because it is dependant on the logic of applicaion. Thus this tier should be separated from other tiers and the process and data coming from business layer is sent to the presentation layer via service interface as the type of XML and the action is performed by Mobile Service. The structure of all layers is shown in Figure 1.

3.2 Presentation Tier

In this Material Information System, The function of UI Components is using UI Process Components sent and those UI Components can be one of type of mobile page or PDA or window applications etc. The function of UI Process Components is to assemble the similar type of input process like multi-button click, similar event processing and then transfer the requests of UI Components to Business Tier. The structure of UI Process Components is shown in Figure 2. UIPressWD Component is used to get the information of goods in warehousing and delivery after calling appropriate method of Service Interface.

UIPressInfo Component is used to get the basic information of product and is also used to update the transaction after calling appropriate method of Service Interface.

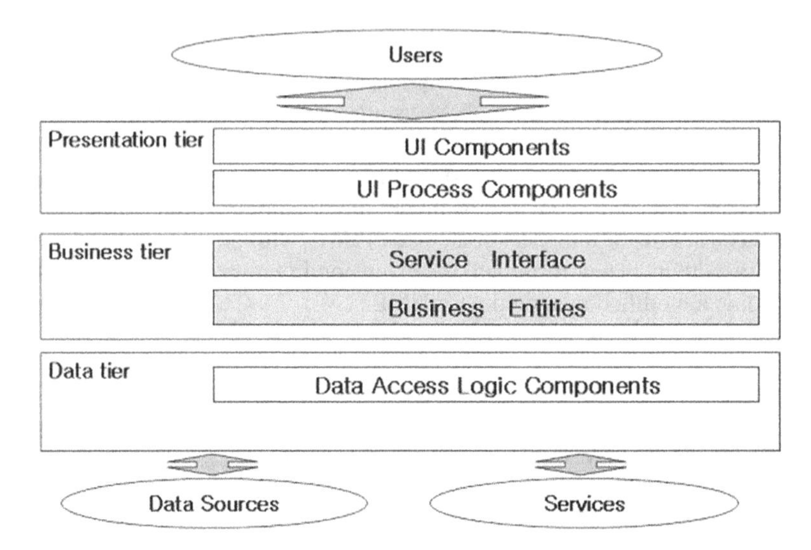

Fig. 1 Basic structure of Mobile component layers

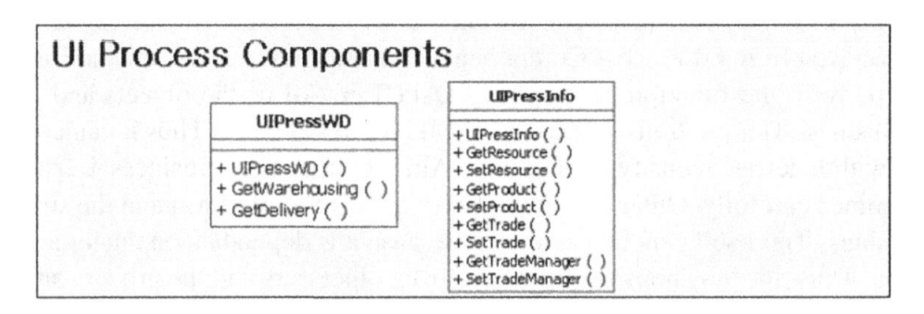

Fig. 2 Software architecture reference models

3.3 Business Tier

The function of Service Interface Component is to transform the business logic into service and mobile service will be included for the function. All the various type of clients request the task to be needed to the exposed service. However the requested task is sent back to the client after passing thru business logic and also transformed into XML type results.

The structure of Service Interfaces and Business Entities is shown in Figure 3. ServiceInfo component manages basic information of the products and ServiceWarehousingDelive-ryM component is used to manage the delivery information. The function of ServiceCommand component is to manage of ordering information and ServiceResourceM component is used to manage stock information. ServicePlan component is used to manage requirement plan and ServiceCommon component manages the material code. All the component in Service Interface access the database in Data tier and return the information to the UIPress components

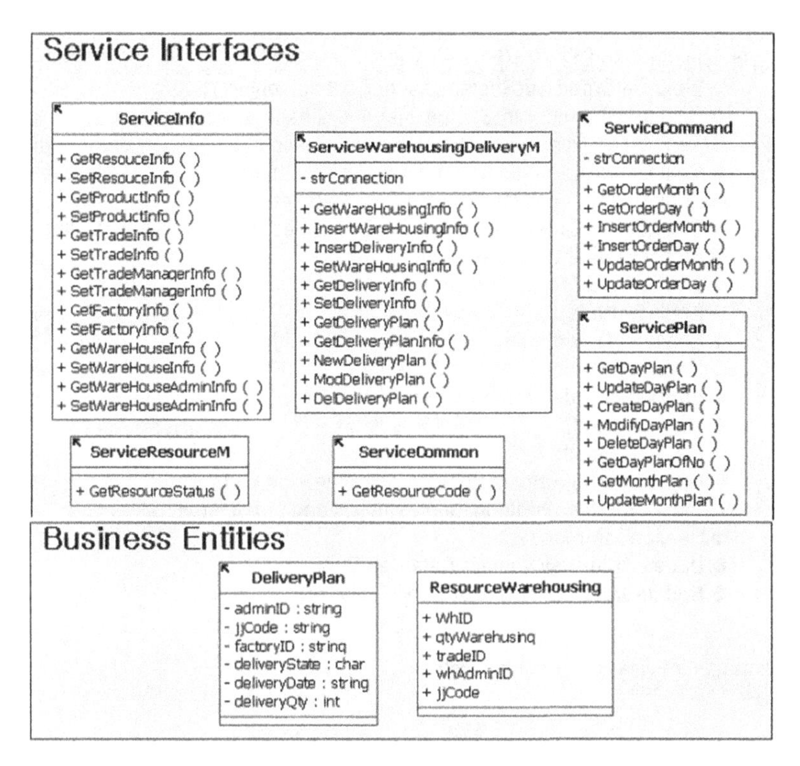

Fig. 3 Component structure of Business Tier

as the type of XML. If there are some entities which depend upon the characteristic of the application the data is transformed into the type of Business Entity object like DeliveryPlan or ResourceWareHousing object.

3.4 Data Tier

During the progress of Business Tier there are many cases to access the data storage. The data and structure are also changed frequently depending upon the environment. It is efficient to construct data access logics separating from others. It helps to enforce the system maintainability [5]. The example of data access logic components in the material management application is shown in Figure 4.

This data access logic component is used in service interface components as mentioned previously.

3.5 Relationship among Components in Each Tier

The concept of the relationship among components in each tier is shown in Figure 5. If one of application in UI Component causes specific event like

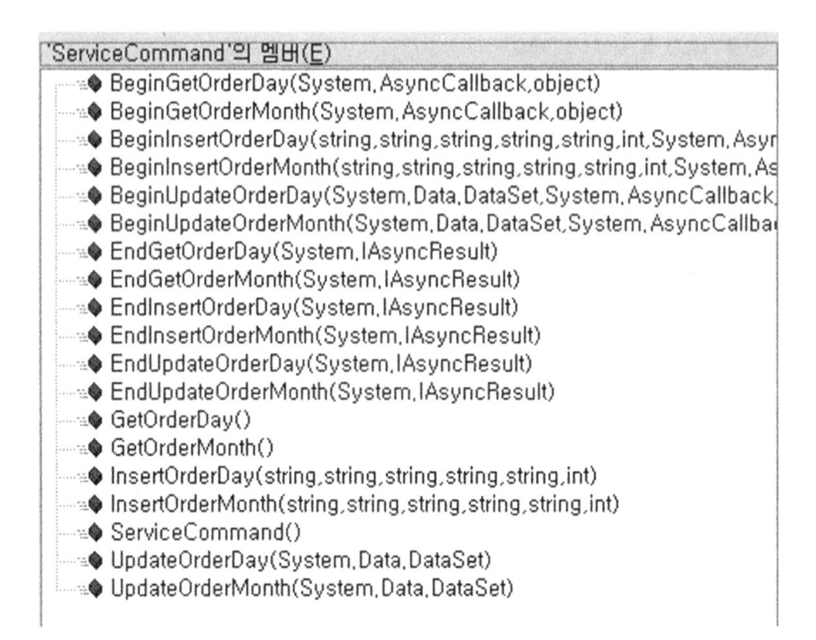

Fig. 4 Example of data access components

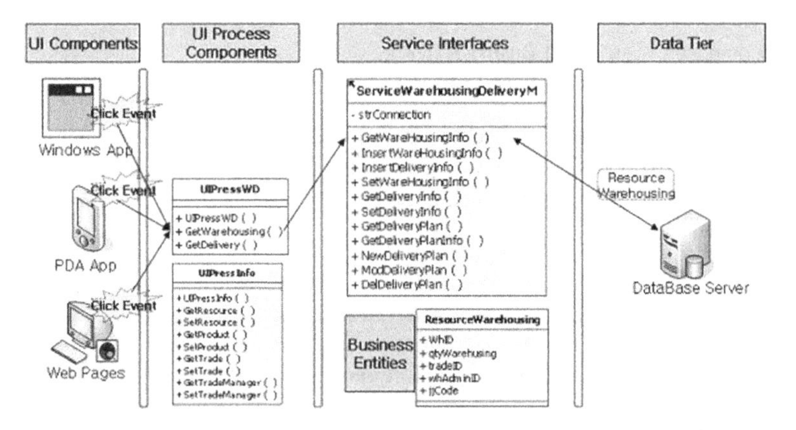

Fig. 5 Relationship among components in each tier

button-click then it calls some method in UI Process Component or it calls some method of Service Interface depending upon the characteristic of the application requirements. The component in Service Interface that is requested from UI Component access the Data Tier and return the results after getting the required information from database.

4 Implementation

As I mentioned in the design phase, implementation can be proceeded as follows.

① : Various type of clients which belong to the UI Components request to UI Process Components
② : UI Process Component calls some method in Service Interface tier for the requested process
③ : One of the component in Business Entity which was called by UI Process Component creates the object to use after accessing the data storage in Data Tier
④ : It returns the object information to the Process Component by the Service Interface as the type of XML.
⑤ : Each client receive the XML typed object and then support the information to the user after parsing and tuning to their own platform environments.

The provision stage before showing the result to the clients are shown in Figure 6 and Figure 7. Down load form of RFID driver and application software in mobile environment is shown in Figure 6 and normally it is used by workers in warehouse. Generally, the intellectual level of workers in warehouse is lower level comparing with others. Thus if there are changes in the software or version, it may be difficult to handle to them. However, this implementation is so helpful for them because they can operate themselves easily in spite of the changes. What they have to do is just following the menu of the device without concerning the type of device and changes.

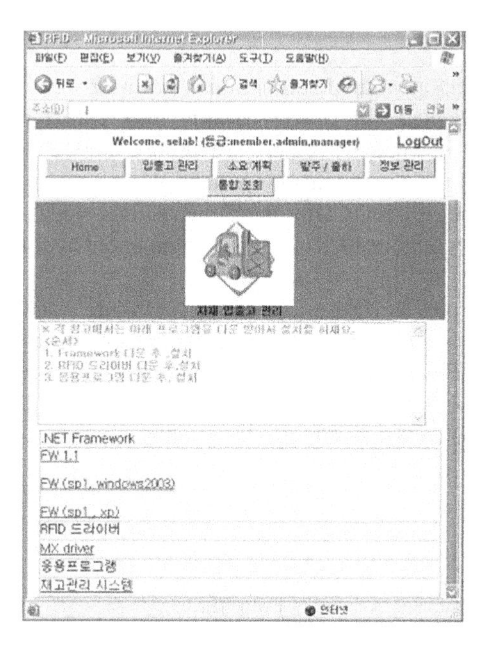

Fig. 6 Down load form of RFID driver and application software in Mobile

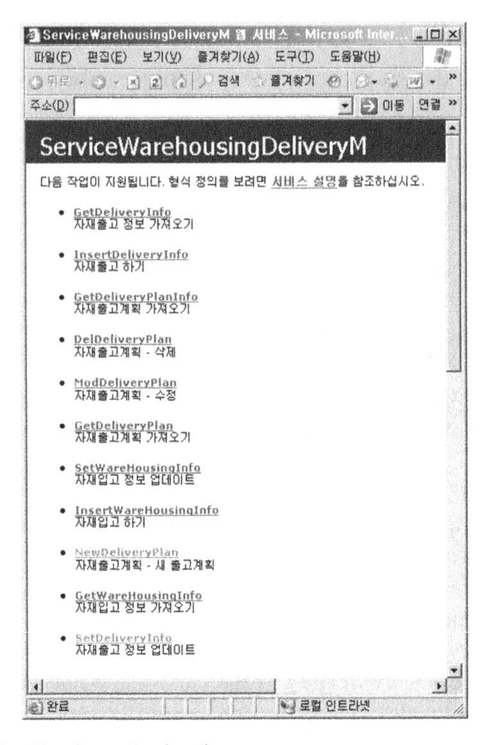

Fig. 7 Component Application selection form

The picture of component application menu that user want to process is shown in Figure 7 and this is the result of transforming into XML typed form. The name of component shown in Figure 7 is ServiceWarehousingDeliveryM and the client can select one of applications in that component. We can see the application results for each different type of clients after these provisional stage and these results are shown from Figure 8 to Figure 11. These are implemented in different types of client system.

We can see the form of RFID implementation in Figure 8 and it is used by worker in warehouse. Window and mobile implementations on desktop are shown in Figure 9 and Figure 10. Finally Mobile device implementation is shown in Figure 10. As we have seen several implementation pictures the designed components can be communicated among each different type of clients.

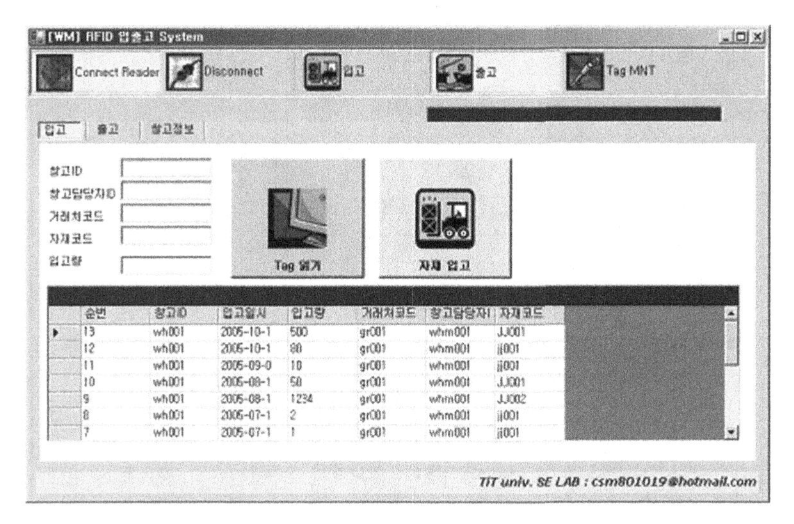

Fig. 8 Component Application selection form

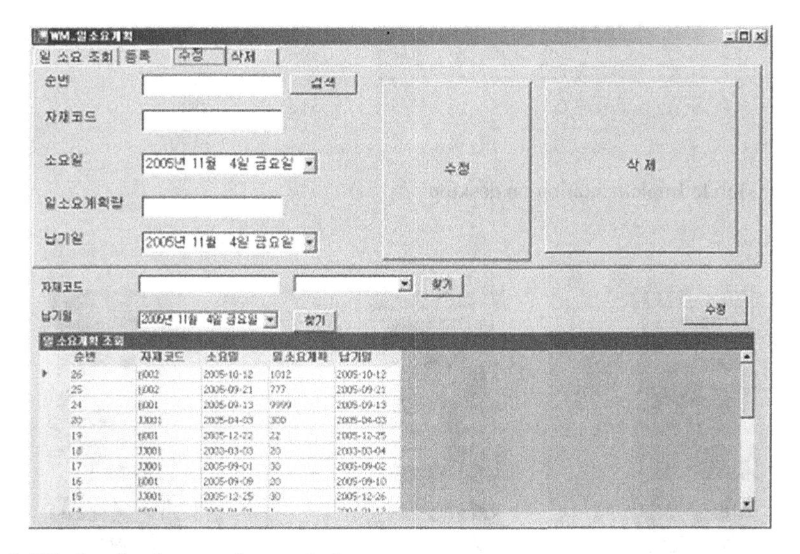

Fig. 9 Window Implementation on desktop

Fig. 10 Mobile Implementation on desktop

Fig. 11 Mobile Implementation

5 Conclusion

In this paper, we suggests the methodology which communicate among components objects under the heterogeneous environments and got the implementation results after design the layered component structure. The trends of mobile service and layered architecture technology was surveyed and utilized as important design factors. We could find the implementation has good results especially in the sense of interoperability. These kind of approach could be so efficient considering current and future application like ubiquitous system because the trend of business environment becomes more complicated and needs to be interfaced with other system more often than before. Thus modular and reusable component will be used so frequently for the most of application field. It could be more effective if design pattern is included in the architecture design that is currently studying and this research shall be verified in the sense of performance measurement in near future.

References

1. Brown, A.W.: Large-Scale Component Based Development. Prentice Hall (2000)
2. Carmichael, A., Haywood, D.: Better Software Faster. Prentice Hall (2002)
3. Hoque, F.: e-Enterprise: Business Models, Architecture and Components. Cambridge University Press, Cambridge (2000)
4. Park, J.S.: Component-Based e-Business Engineering. In: 4th International Conference on Electronic Commerce Research, Dallas, TX (November 2001)
5. Larsen, G.: Component-Based Enterprise Frameworks. Communications of ACM 43(10) (October 2000)
6. Marinescu, F.: 2001: The Year of Mobile Services. Java Developer's Journal (April 2001)
7. Smith, D., et al.: The Future of Mobile Services: Dynamic Business Mobiles. Gartner Research Note (April 2001)
8. Rumbaugh, J., Blaha, M., Premerlani, W., Eddy, S., Lorensen, W.: Object-Oriented Modelling and Design. Prentice-Hall, Englewood Cliffs (1991)
9. Jacobson, I., Booch, G., Rumbaugh, J.: The Unified Software Development Process. Addison-Wesley, Reading (1999)
10. Turner, M.S.V.: Microsoft Solutions Framework Essentials 2006: Microsoft Corporation (2006)
11. Preece, J., Rogers, Y., Sharp, H.: Interaction Design: Beyond Human-Computer Interaction. John Wiley and Sons, NewYork (2002)
12. Vaskevitch, D.: Client/Server Strategies. IDG Books, San Mateo (1993)
13. Kim, H.-K., Kwon, O.-H.: SCTE: Software Component Testing Environments. In: Gervasi, O., Gavrilova, M.L., Kumar, V., Laganá, A., Lee, H.P., Mun, Y., Taniar, D., Tan, C.J.K. (eds.) ICCSA 2005. LNCS, vol. 3481, pp. 137–146. Springer, Heidelberg (2005)
14. Szyperski, C.: Component Software Beyond Object-Oriented Programming, 2nd edn. Addison-Wesley (2002)

A Optimal Scheduling Strategy for Data-Driven Peer-to-Peer Streaming

Guowei Huang and Zhi Chen

Abstract. In recent years, the data-driven peer-to-peer streaming systems have been extensive deployed in Internet. In these systems, the transfer of media blocks among nodes involves two scheduling issues: each node should request the streaming blocks of interest from its peers (i.e. block request scheduling), on the other hand, it also should decide how to satisfy the requests received from its peers (i.e. block delivery scheduling) given its bandwidth limitations. Intuitively, these two scheduling issues are critical to the performance of data-driven streaming systems. However, most of the work in the literature focused on the block request scheduling issue, and very few concentrated on the latter. Consequently, the performance of the system may be affected seriously due to an unsophisticated scheduling strategy. In this paper, we analytically study the block delivery scheduling problem and model it as an optimization problem based on the satisfaction degrees of nodes and the playback performance of the system. We then propose a scheduling strategy and prove the optimality of the strategy to the optimization problem. Lastly, we illustrate the effectiveness of the proposed strategy by extensive simulation.

Keywords: peer-to-peer, data-driven streaming, block delivery scheduling.

1 Introduction

Video streaming is part of the basic service that we expect in the current Internet. There has been a number of studies on how to provide streaming service using the

Guowei Huang
Computer College, Shenzhen Institute of Information Technology, Shenzhen, China
e-mail: `huanggw@sziit.edu.cn`

Zhi Chen
School of Computer Science and Software Engineering, Tianjin Polytechnic University, Tianjin, China
e-mail: `mufchen@gmail.com`

R. Lee (Ed.): Computer and Information Science 2012, SCI 429, pp. 165–180.
springerlink.com

client-server architecture and how to engineer streaming servers so as to provide the quality-of-service guarantees [1]. In recent years, the attention is on how to provide a scalable streaming service to a large population of viewers. To this end, IP multicast was proposed so that the server only needs to send a copy of video file and routers along the distribution network will relay all packets to different end users. However, due to security and deployment issues [2], IP multicast has not been widely deployed. Instead, people are using application layer multicast to deliver the video files to users.

Peer-to-peer (P2P) system is considered one form of application layer multicast. In particular, the data-driven streaming built on P2P systems is shown to exhibit good scalability property: the service rate of the data-driven based P2P streaming system is proportional to the number of users. Therefore, Internet has witnessed a rapid growth in the deployment of data-driven based peer-to-peer streaming systems (such as UUSee [3], PPLive [4], Coolstreaming [5], etc.) in recent years. Measurements on these systems report over 100,000 concurrent users on a single channel for PPLive [6] and over 25,000 concurrent users on a single channel for CoolStreaming [7].

The basic idea of data-driven streaming protocol is very simple: the media content is divided into *blocks* and the transfer progress of blocks among nodes can be characterized by a block scheduling issue: every node announces what blocks it currently holds to its neighbors, then each node requests the blocks of interest from its neighbors according to their announcements. Concretely, the block scheduling issue can be further split into two components: block request scheduling and block delivery scheduling. In the block request scheduling, every node should decide from which neighbor to ask for which blocks, based on the block announcements received from its neighbors. In the block delivery scheduling, on the other hand, every node should decide how to satisfy the block requests which it receives from its neighbors, given its bandwidth limitations. Intuitively, these two scheduling issues are critical to the performance of data-driven streaming system.

To improve the performance of data-driven streaming system, most of the recent papers concentrated on the block request scheduling issue [8, 9, 10, 11, 12, 13, 14, 15]. However, few research efforts have been made with respect to the block delivery scheduling yet. Most of existing data-driven streaming applications [3, 4] adopt the First Come First Serve (FCFS) strategy in the block delivery scheduling: every node queues the requests that it received in increasing order of their arriving time; and it preemptively satisfies the requests arriving earlier.

However, the shortcomings of FCFS-based scheduling strategy are obvious: the feature of FCFS may cause large response time or even starvation for some requesters when the request queue is long, which leads to bad experiences of these requesters. Furthermore, FCFS-based scheduling strategy ignores the potential influence of a desired block of a node on the streaming playback quality observed by that node. As consequence, the strategy may result in the degradation of the playback performance of the streaming system. For a P2P streaming service provider, it is vital to achieve both the high satisfaction degrees of users and the high streaming playback performance at the same time. Therefore, a more sophisticated scheduling strategy should take these two factors into account.

Based on the above considerations, we present our analytical model and corresponding solution to tackle the block delivery scheduling issue in data-driven streaming. First, we introduce the concepts of *utility* and *playback quality weight*, to represent the importance of a block that a node has interest in to the individual satisfaction degree of that node and the playback performance of the system, respectively. Second, we model the block delivery scheduling issue as an optimization problem and propose a greedy-based scheduling strategy in the light of the model. The strategy has linearithmic time complexity and we show that the output of the strategy is optimal to the model. Simulation results shows the effectiveness of our scheduling strategy in improving the satisfaction degrees of users as well as the playback performance of the streaming system.

The remainder of this paper is organized as follows: The related work of this paper is briefly reviewed in Sect. 2. Section 3 presents our model for the block delivery scheduling issue, and proposes a greedy-based strategy based on the model. In Sect. 4, we conduct simulations to evaluate the effectiveness of the proposed strategy. Finally, we conclude this paper and give our future work in Sect. 5.

2 Related Work

Due to the importance of the block request scheduling to the performance of data-driven streaming protocol, many research efforts have been made on this issue. Chainsaw [8] uses a pure random scheduling strategy: each node assigns each desired block randomly to a neighbor which holds that block, while DONet [5] adopts the Local Rarest First (LRF) method: a block that has the minimum owners among the neighbors will be requested first. Reference [9] proposes a probabilistic scheduling strategy: the probability that a desired block will be requested in the scheduling is inversely proportional to the playback distance between that block and the playback point of the requester. PPLIVE [4] adopts a mixed scheduling strategy: giving the first priority to the sequential property of the desired block, and then the rarity property. In [10], the desired blocks of a node are categorized into three classes according to the emergent degrees of these blocks to the node, and the scheduling gives the first priority to the most emergent blocks. Reference [11] deals with the neighbor selection issue that a node should ask which neighbor for its desired blocks. It suggests that the request of a node should be sent to the neighbor with the shortest requesting queue, in order to guarantee the load balancing in the neighborhood of the node.

There are also many researches which focus on analyzing the impact of the block requesting scheduling strategy on the performance of the data-driven streaming protocol. Reference [13] analyzes the impact of different scheduling strategies (including LRF strategy and sequential strategy) on the sequential progress and the startup delay of the streaming system, based on the fluid model. References [9, 15] analyze the impact of different request scheduling strategies on the throughput performance of the streaming system, based on the concept of entropy. Reference [14] models the block request scheduling issue as a classical min-cost network flow problem, and proposes both the global optimal scheduling scheme and distributed heuristic algorithm to optimize the system throughput.

3 Block Delivery Scheduling: Problem Statement and Formulation

3.1 Block Delivery Scheduling Problem

First of all, we briefly review some characteristics of the data-driven streaming protocol here. The idea of data-driven P2P streaming protocol is very similar to that of Bit-Torrent protocol [16]. In such protocol, media content are partitioned into blocks with the equal size, each of which has a unique sequence number. Every node periodically announces what blocks it holds to all its neighbors. Due to the announcement of the neighbors, each node periodically sends requests to its neighbors for the desired blocks. Furthermore, each node plays the media blocks one by one at the speed of streaming rate, according to the sequence numbers of blocks. Therefore, the streaming playback quality of a node relies on the delivery timeliness of its desired blocks: the block that a node is interested in should be delivered to that node before the playback deadline of the block. Otherwise, a block delayed beyond the deadline will lead to the degradation of the playback quality observed by the requester.

In data-driven streaming protocol, each node determines how to satisfy the requests received from its neighbors, according to the block delivery scheduling strategy that it adopts. Therefore, the block delivery scheduling strategy plays an important role in affecting the playback performance of the streaming system as well as the satisfaction degrees of nodes with the block delivery service:

- From the perspective of a block requester, it always expects that it can receive the desired blocks from its neighbors as many as possible. If a requester, on the other hand, finds that it often obtains few or even none of desired blocks from its neighbors, it will be disappointed with the block delivery and has no interest in staying in the system any more.
- Intuitively, to guarantee the playback quality of the streaming system, the blocks in danger of being delayed beyond their deadlines should be given more priority in the block delivery scheduling.

However, in the FCFS-based scheduling strategy, whether a block request should be satisfied or not by the scheduling node is entirely determined by the arriving time of the request. Consequently, the scheduling results of the strategy may be suboptimal from the views of the satisfaction degrees of requesters and the playback quality of the streaming system.

Figure 1 gives an intuitive example of the block delivery scheduling problem. The blocks close to node N_1 and N_2 illustrate the status in the block buffers of these two nodes: N_1 holds block 3, 4, 5 and N_2 holds blocks 4, 6, 7, 8, respectively. Furthermore, the current playback points of N_1 and N_2 are block 3 and block 4, respectively. Node N_s is a common neighbor of N_1 and N_2, and it holds all the desired blocks of these two nodes. The upload bandwidth of N_s allows it to upload only two blocks in one delivery scheduling period. Suppose that N_1 decides to ask N_s for blocks 6, 7, and N_2 decides to get block 5 from N_s. Suppose a scenario in which the requests of N_1 arrive at N_s a little earlier than that of N_2. According to the FCFS-based strategy, N_s will only satisfy the requests of N_1, while the request of N_2 has to be delayed due to the exhaustion of upload bandwidth of N_s. However, such scheduling strategy has the following shortcomings:

- First, the scheduling strategy leads to the extreme in terms of the satisfaction degrees of N_1 and N_2: the satisfaction degree of N_1 is maximized, since all the requests of N_1 are satisfied. On the other hand, N_2 is totally dissatisfied with the scheduling, since the scheduling brings it into the starvation status.
- Second, the scheduling strategy may cause the degradation in the playback quality of N_2. Since block 5 is the next playback block of N_2, the above scheduling result will cause block 5 in danger of being delayed beyond its playback deadline with high probability.

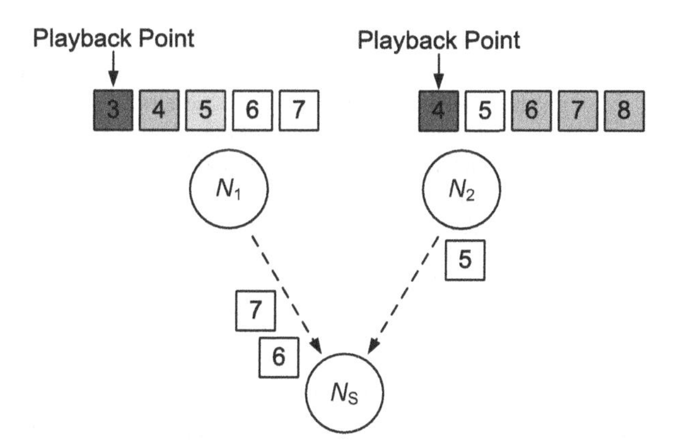

Fig. 1 Illustration of block delivery scheduling problem

Compared with the above scheduling strategy, a better scheduling strategy is that N_s satisfies the request of N_1 for block 6 and the request of N_2 for block 5, respectively: such scheduling strategy maximizes the satisfaction degree of N_2, while avoiding to bring starvation to N_1 and decreasing the degradation probability for the playback quality observed by N_2.

3.2 Model and Solution

In this section, we discuss how a scheduling node, say N_s, implements a block delivery scheduling strategy to distribute its bandwidth resource among all the requests received from its neighbors. The notations that we use in the following analysis are shown in Table 1. To simplify the analysis, we assume that the block delivery scheduling is performed in a periodic manner by each node: at the beginning of its scheduling period, each node decides how to satisfy the block requests which it received from its neighbors recently.

Table 1 Notations

Notations	Descriptions
u_{ij}	The utility weight of a desired block j of node N_i
p_{ij}	The playback quality weight of a desired block j of node N_i
C_i	The current playback point of node N_i
H_{ij}	The number of node N_i's neighbors that hold block j
S	The set of the nodes which send requests to node N_s
R_i	The set of desired blocks which node N_i wants to get from node N_s
R	The set of desired blocks which the nodes in S want to get from N_s. Any two elements in R can be differentiated by their sequence numbers and the identities of their requesters.
B_i	The available end-to-end bandwidth from node N_s to node N_i
B	The available upload bandwidth of node N_s
$\|M\|$	The size of a set M
t	The block delivery scheduling period
b	The block size
r	The streaming playback rate

We start with analyzing and representing the influence of block delivery scheduling strategy adopted by N_s on the satisfaction degrees and the playback qualities of its requesters, respectively. Based on the analysis, we then model the block delivery scheduling issue as an optimization problem. In the light of the model, we propose our scheduling strategy so as to improve the satisfaction degrees of nodes and the playback quality of streaming system at the same time.

3.2.1 Satisfaction Degree of Nodes

Suppose node N_i is a requester of N_s. Let R_i denote the set of desired blocks that N_i wants to obtain from N_s. Let $x_i = \{x_{ij} | j \in R_i\}$ denote a set of decision variables corresponding to R_i. Concretely, each elements of x_i, say x_{ij} $(j \in R_i)$, is a decision variable denoting whether the desired block j of N_i will be delivered by N_s in the scheduling: $x_{ij} = 1$ implies that N_i will obtain block j from N_s, vice versa.

From the perspective of requester N_i, its satisfaction degree with the block delivery service received from N_s is determined by how its desired blocks are delivered in the scheduling, that is, the values of x_{ij}. Since different scheduling strategies adopted by N_s may result in different values of x_{ij}, different strategies may make a great difference in the satisfaction degree of N_i. To overcome the problem, we use the concept of *utility* to represent the degree of satisfaction of a requester. Given R_i and x_i, the utility of N_i denoted by $U_i(R_i, x_i)$ is defined as follows:

$$U_i(R_i, x_i) = \sum_{j \in R_i} u_{ij} x_{ij} , \tag{1}$$

where u_{ij} ($u_{ij} \geq 0$) is the *utility weight* of block j to N_i, which reflects the importance of block j to the satisfaction degree of N_i, compared with other desired blocks in R_i.

The utility function we consider in this work satisfies the following assumptions:

- $U_i(R_i, x_i) = 0$, if all the elements of x_i equals to 0.
- $U_i(R_i, x_i) = 1$, if all the elements of x_i equals to 1.

The justifications of the above assumptions are as follows. First, if a requester does not get any desired blocks from N_s in the scheduling, the requester will not satisfy with the block delivery service provided by N_s at all. On the other hand, if all the requests of a requester are satisfied by N_s, the satisfaction degree of the requester should reach the maximum. Second, the above assumptions normalize the utility of all nodes so that we can compare the degrees of satisfaction of different nodes.

3.2.2 Playback Performance

As mentioned before, from the perspective of a node, the playback performance provided by the streaming system is determined by whether it can receive its desired blocks from the system in time. Intuitively, each scheduling node N_s should preemptively deliver the blocks which are more significant to the playback qualities of the requesters, to guarantee the playback performance of the system. Therefore, we introduce the concept of *playback quality weight* of a desired block, to represent the significance of the block to the playback quality of the requester. Two properties of a desired block are considered in the definition of playback quality weight: the emergent property and the rarity property.

First, the emergent degree of a desired block is one of the important factors which influence the delivery timeliness of the block from the view of the requester: the more emergent the desired block is (i.e. the closer the block is to the playback point of the requester), the higher the probability that the block will be delayed beyond its deadline is, which implies that the desired block is more important to guarantee the playback quality of the requester.

Second, the delivery timeliness of a desired block for the requester is also closely related to the rarity property of the block. Intuitively, the more copies of the block there are in the requester's neighborhood, the more likely the requester can get the block from its neighbors. Therefore, the requests for the blocks with few holders should be satisfied preemptively. Furthermore, the previous empirical study [17]has shown that “rarest-first” is very efficient strategy in data dissemination.

Consequently, the playback quality weight p_{ij} of a desired block j ($j \in R_i$) of node N_i is defined as:

$$p_{ij} = \alpha P_E(j - C_i) + (1 - \alpha) P_R(H_{ij}) \qquad (0 \leq \alpha \leq 1)\,. \tag{2}$$

The function $P_E(x)$ in the first item represents the emergency property and $j - C_i$ represents the remaining time of block j till its playback deadline. The function $P_R(x)$ in the second item represents the rarity property and H_{ij} is the number of node i's neighbors that hold block j. Both functions should be monotonously non-increasing. For emergency property, we hope that each desired block has different priority in terms of their remaining time till playback deadline, so we set $P_E(x) = 10^{9-b*x/r}$, when $b * x/r \leq 9$; otherwise, $P_E(x) = 1$. For Rarity property, we define $P_R(x) = 10^{9-x}$, when $x = 1, \ldots, 8$; otherwise, $P_R(x) = 1$.

3.2.3 Model

Our goal is to improve both the satisfaction degrees of nodes and the playback performance of streaming system at the same time. To achieve this goal, naturally, each scheduling node N_s should give priority to the desired blocks with high utility weight or (and) high playback quality weight, given its bandwidth limitations. Mathematically, we model such block delivery scheduling thought as the following optimization problem:

$$\begin{aligned}
&\max \sum_{N_i \in S} \sum_{j \in R_i} u_{ij} p_{ij} x_{ij} \\
&s.t. \\
&(a) \quad \sum_{j \in R_i} x_{ij} \leq B_i t/b \qquad (N_i \in S) \\
&(b) \quad \sum_{N_i \in S} \sum_{j \in R_i} x_{ij} \leq Bt/b
\end{aligned} \tag{3}$$

Constraints (a) and (b) ensure the scheduling satisfy the bandwidth capacity limitations of the scheduling node and the requesting nodes, respectively.

We call the above optimization problem a Block Delivery Scheduling with Utility and Playback quality (BDS-UP for short). It is important to point out that BDS-UP strategy implies that the delivery priority of a desired block in the scheduling is equivalent to the product of its utility weight and playback quality weight.

3.2.4 Solution

In this subsection, we propose a greedy-based scheduling strategy for the above optimization problem BDS-UP and prove the optimality of the output of the strategy. The pseudo-code of the scheduling strategy for a scheduling node N_s is shown in Fig. 2.

```
1.   for (each requester N_i in S)
2.       for (each block j in R_i)
3.           compute the delivery priority of the block
4.       end for
5.   end for
6.   Sort the desired blocks in R, in descending order of delivery priority
7.   for m = 1 to |R|
8.      if the mth block in R is the block j requested by node N_i then
9.         if (Bt − b) > 0 and (B_i t − b) > 0 then
10.            B ← (Bt − b)/t
11.            B_i ← (B_i t − b)/t
12.            x_ij ← 1
13.        else
14.            x_ij ← 0
15.        end if
16.     end if
17.  end for
```

Fig. 2 Pseudo-code of the greedy-based scheduling strategy of node N_s for BDS-UP

According to the pseudo-code, the scheduling node first sorts the blocks in descending order of delivery priority (Line 6). Then the scheduling node chooses the block with the highest priority currently in each scheduling step (Lines 7–17), while guaranteeing that the available bandwidth capacity limitations are satisfied (Line 9). Note that the strategy has a time complexity of $O(nlog(n))$, where n is the size of R. Therefore, the scheduling can be performed efficiently.

As shown in Fig. 2, to execute the strategy, the scheduling node N_s has to collect enough information from its requesters, such as the information about the emergency and rarity properties, the utility weights of the blocks desired by the requesters, the end-to-end available bandwidth from the scheduling node to each requester. To achieve this goal, we require that each requester N_i should provide the necessary information to the scheduling node N_s along with its requests. The information includes the current playback point of requester N_i, the utility weights of its desired blocks, the distributions of its desired blocks in its neighborhood and the end-to-end bandwidth from N_s to N_i.

However, it is difficult for a requester N_i to know the exact end-to-end available bandwidth from the scheduling node N_s, that is, B_i. For simplicity, we use a heuristic way proposed by [14] for each requester to estimate the maximum rate at which the scheduling node can send blocks to it. We let $b_i(k)$ denote the number of blocks received by N_i from its neighbor N_s in the kth period. When N_i asks N_s for its desired blocks, N_i use the average traffic received by N_i in the previous K periods to estimate B_i in the $(k+1)$th period:

$$B_i = (1+\theta) \sum_{m=k-K+1}^{k} b_i(m)/K\,, \tag{4}$$

where θ (named aggressive coefficient) is a small positive constant for the bandwidth estimation. The purpose of the aggressive coefficient is to explore the possibility that the end-to-end bandwidth from N_s to N_i may increase recently.

The following lemma and theorem prove that the output of the proposed strategy is an optimal solution for the optimization problem BDS-UP.

Lemma 1. *Let $x = \{x_{ij} \,|N_i \in S, j \in R_i\}$ be the output of the proposed strategy and $x' = \{x'_{ij} \,|N_i \in S, j \in R_i\}$ be an optimal solution of the optimization problem BDS-UP, then we have*

$$\sum_{N_i \in S} \sum_{j \in R_i} x_{ij} = \sum_{N_i \in S} \sum_{j \in R_i} x'_{ij} \,. \tag{5}$$

Proof. Assume that

$$\sum_{N_i \in S} \sum_{j \in R_i} x_{ij} < \sum_{N_i \in S} \sum_{j \in R_i} x'_{ij} \,, \tag{6}$$

then we can get:

- There must exist a node N_i such that $\sum_{j \in R_i} x_{ij} < \sum_{j \in R_i} x'_{ij}$. It implies that there is at least one desired block of node N_i which is not satisfied after the execution of the strategy, due to the exhaustion of the available upload bandwidth of the scheduling node N_s.
- According to the constraint (b) of BDS-UP, the above assumption implies that there is still enough available upload bandwidth of N_s to transfer one block after the execution of the strategy.

Therefore, we have a contradiction. Let us further assume that

$$\sum_{N_i \in S} \sum_{j \in R_i} x_{ij} > \sum_{N_i \in S} \sum_{j \in R_i} x'_{ij} \,, \tag{7}$$

then we can get:

- There must exist a node N_i such that $\sum_{j \in R_i} x_{ij} > \sum_{j \in R_i} x'_{ij}$. It implies that there exists at least one desired block of N_i is not satisfied in the optimal solution, due to the exhaustion of the available upload bandwidth of the scheduling node N_s.
- According to the constraint (b), the assumption implies that there is still enough available upload bandwidth of N_s to transfer one block in the optimal solution.

Therefore, $x' = \{x'_{ij} \,|N_i \in S, j \in R_i\}$ is not the optimal solution.We have a contradiction. □

Lemma 1 implies that the number of desired blocks of requesters delivered by the scheduling node in the output of the proposed strategy equals to that in the optimal solution.

Theorem 1. *The output of the proposed strategy is an optimal solution of the optimization problem BDS-UP.*

Proof. Suppose node N_s is the scheduling node. Let R denote the set of desired blocks which the neighbors of N_s want to get from N_s and D denote the size of R. Let $w = \{w_1, w_2, \ldots, w_D\}$ be the set of the delivery priorities of the blocks in R, and w is sorted in descending order. Let $x = \{x_{w_1}, x_{w_2}, \ldots, x_{w_D}\}$ be the output of the proposed strategy corresponding to w and $x' = \{x'_{w_1}, x'_{w_2}, \ldots, x'_{w_D}\}$ be an optimal solution corresponding to w, respectively.

According to Lemma 3.2.4, the number of the elements which is equal to 1 in the set x is the same as that in x'. Let n be the number of the elements which equals to 1 in x (or x'). Let $f_g(i)$ be the function which returns the index of the ith $(1 \leq i \leq n)$ element which equals to 1 in the set x. Let $f_o(i)$ be the function which returns the index of the ith $(1 \leq i \leq n)$ element which equals to 1 in the set x'.

To prove the theorem, we first prove that for any i $(1 \leq i \leq n)$, we have

$$w_{f_g(i)} \geq w_{f_o(i)} \,. \tag{8}$$

Assume that there exists an integer m $(1 \leq m \leq n)$ such that $w_{f_g(m)} < w_{f_o(m)}$. As w is sorted in the descending order, then the assumption implies that $f_g(m) > f_o(m)$. Let $M_{f_o(m)}$ be the set of the requesters corresponding to $\{w_1, w_2, \ldots, w_{f_o(m)}\}$. Therefore, there must exist a node $N_t \in M_{f_o(m)}$ such that the number of the requests of node N_t which is satisfied in $\{x_{w_1}, x_{w_2}, \ldots, x_{w_{fo(m)}}\}$, is smaller than the number of the requests of node t which is satisfied in $\{x'_{w_1}, x'_{w_2}, \ldots, x'_{w_{fo(m)}}\}$. This observation implies that as least one desired block of N_t, whose delivery priority is larger than $w_{f_g(m)}$, would be delivered in $\{x_{w_1}, x_{w_2}, \ldots, x_{w_{fo(m)}}\}$ in the proposed strategy instead of the block corresponding to $w_{f_g(m)}$. Therefore, we have a contradiction. As a result, we get that for any i $(1 \leq i \leq n)$,

$$w_{f_g(i)} \geq w_{f_o(i)} \,. \tag{9}$$

As a consequence, we have that

$$\sum_{1 \leq i \leq n} w_{f_g(i)} x_{w_{f_g(i)}} \geq \sum_{1 \leq i \leq n} w_{f_o(i)} x'_{w_{f_o(i)}} \,, \tag{10}$$

which implies that the solution of the proposed strategy is the optimal solution to the optimization problem. □

4 Performance Evaluation

4.1 Experiment Setup

In our experiments, we implement a discrete event-driven simulator to evaluate the performance of our scheduling strategy. For a fair comparison, all the experiments use the same simple algorithm for overlay construction: each node independently selects its neighbors randomly so that a random graph is organized. Moreover, to evaluate the playback performance, we define the concept of *timeliness ratio* to represent the number of blocks that arrive at each node before their playback deadlines over the total number of blocks. The average timeliness ratio indicates the playback quality of the whole system. On the other hand, we use the average utility which a requester can get from one block delivery scheduling period to evaluate the satisfaction degree of a node. In our experiments, all of nodes adopt the strategy proposed by [4] in block request scheduling. The parameters and default values used in the experiments are summarized in Table 2.

Table 2 Experiment parameters

Parameters	Default values
Number of nodes	300
Streaming playback rate r	400 kbps
Scheduling period t	0.8 second
Block size b	1250 bytes
Weight coefficient α	0.4
Aggressive coefficient θ	0.5
Number of neighbors of each node	14
Upload bandwidth of the streaming source	2 Mbps

For node outbound and inbound bandwidth, we adopt the measurement results derived from actual Gnutella nodes in [18]. In our experiments, there are three types of nodes in terms of the bandwidth capacity by default: 20% nodes with outbound bandwidth 128 kbps and inbound bandwidth 784 kbps, 50% nodes with outbound bandwidth 384 kbps and inbound bandwidth 1.5 Mbps, 30% nodes with outbound bandwidth 1 Mbps and inbound bandwidth 3 Mbps. Therefore, the default average outbound bandwidth of nodes in the experiments is about 517 kbps. The reason that we use such a distribution of node's capacity is to account for the situation of node capacity in the real: most of users in Internet are DSL/Cable users nowadays.

In our experiments, we use a simple configuration for the utility weights of desired blocks of a requester N_i, that is, all the desired blocks of N_i have the same utility weights: $u_{ij} = 1/|R_i|$, $(j \in R_i)$.

Besides FCFS-based strategy, we also implement two simple strategies in our experiments for the sake of comparison:

The first strategy is called Block Delivery Scheduling with Utility (BDS-U for short). The feature of this strategy is that it only considers the influences of desired blocks in the utility of nodes, regardless of the influences of desired blocks on the playback performance. Therefore, The BDS-U strategy can be viewed as a specialization of BDS-UP strategy by setting $p_{ij} = 1$ for all blocks.

The second strategy is called Block Delivery Scheduling with Playback quality (BDS-P for short). Contrary to BDS-U, this strategy only considers the influences of desired blocks on the playback performance. Therefore, The BDS-U strategy can be viewed as a specialization of BDS-UP strategy by setting $u_{ij} = 1$ for all blocks.

4.2 Impact of the Weight Coefficient α

In this experiment, we record the performance of BDS-UP strategy by varying the playback quality weight coefficient. Our goal is to demonstrate the impact of the weight coefficient α on the performance of BDS-UP strategy and to find the reasonable value for α.

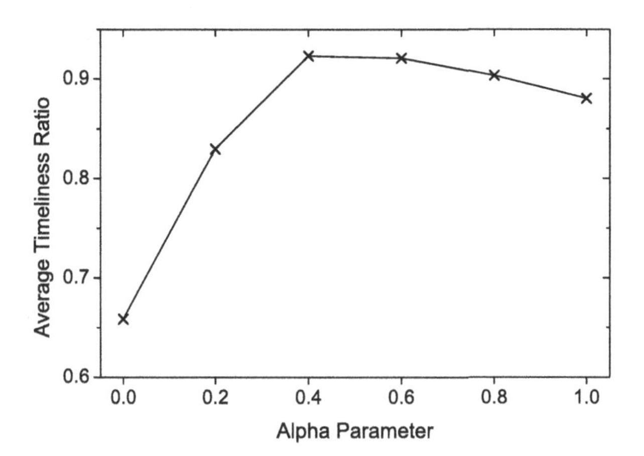

Fig. 3 Timeliness ratio for different value of α

Figure 3 shows the timeliness ratio under different value of $\alpha \in [0,1]$. We observe a steep jump when the value of α changes from 0.0 to 0.2. It implies that the rarity property of blocks is more important to improve the playback quality of nodes compared to emergency property. Some similar results are also obtained in [19]. An explanation is that requesting more emergent blocks will impact the diversity of blocks in the streaming system, which leads to the decrease of the chance that a node can exchange blocks with its neighbors [17]. As shown in Fig. 3, BDS-UP strategy achieves the highest timeliness ratio when α is equal to 0.4. Therefore, we set $\alpha = 0.4$ in all the other experiments.

4.3 Varying the Streaming Rate

The objective of this experiment is to evaluate the performance of each strategy by varying the streaming rate of system. The experiment results are shown in Fig. 4. We see that when the streaming rate is 100 kbps, all of four strategies achieve high timeliness ratio and high average utility at the same time. The reason is that the average bandwidth capacity of streaming system is so sufficient that all the requests of nodes can be satisfied with high probability. However, as the streaming rate increases, the performances of each strategy shows a descending trend, due to the deficit of bandwidth capacity. In particular, the playback performance of BDS-U strategy and FCFS-based strategy goes down fast with the increase of streaming rate. On the other hand,the playback performance of BDS-UP strategy and BDS-P strategy still keeps in a high level. The advantage of BDS-UP strategy over BDS-P strategy lies in the satisfaction degree of nodes: at the rate of 450 kbps, BDS-UP strategy outperforms BDS-P strategy by gains 12% in terms of average utility.

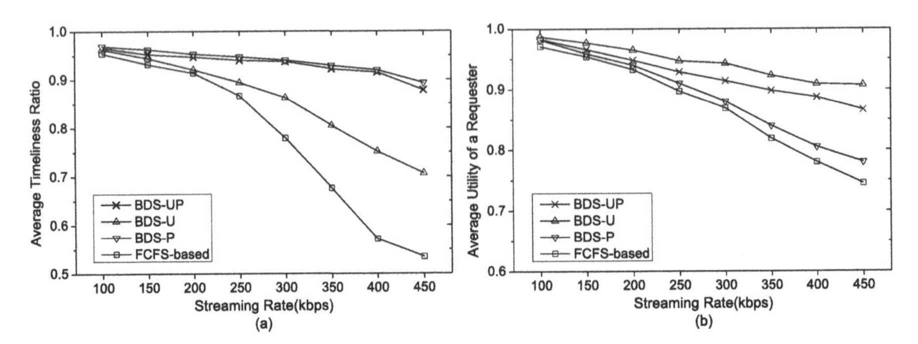

Fig. 4 (**a**) Timeliness ratio and (**b**) average utility for different streaming rates

4.4 The Impact of Nodes with High Bandwidth Capacity

In this experiment, we study the impact of nodes with high capacity on the performance of the streaming system. We change the experiment configuration as follows: 10% nodes with inbound bandwidth 10 Mbps and outbound bandwidth 5 Mbps, which correspond to Ethernet users in Internet. The remaining 90% nodes are also classified according to the rule described in Sect. 4.1. The resulting average outbound bandwidth of nodes is about 942 kbps.

As shown in Fig. 5, the performance of each strategy is improved after adding the nodes with high bandwidth capacity, compared with Fig. 4. The reason is that the bandwidth capacity of streaming system gets richer due to the addition of the nodes with high capacity. However, BDS-UP is still significantly outperforms FCFS-based strategy and BDS-U strategy in terms of playback performance as the increasing of streaming rate: at the rate of 450 kbps, BDS-UP strategy outperforms FCFS-based strategy and BDS-U strategy by gains of 29% and 10%, respectively. On the other hand, BDS-UP strategy still keeps significant advantage over BDS-P strategy in terms of satisfaction degree of nodes: the gap between BDS-UP and BDS-P is 10%, at the rate of 450 kbps.

4.5 Varying the Group Size

Figure 6 shows the performance of each strategy with respect to different group sizes. We can observe that the curves of all the four strategies are flat, which implies that the performance of each strategy remains almost the same when the group size increases. This indicates that the group size has little impact on the performance of data-driven P2P streaming system.

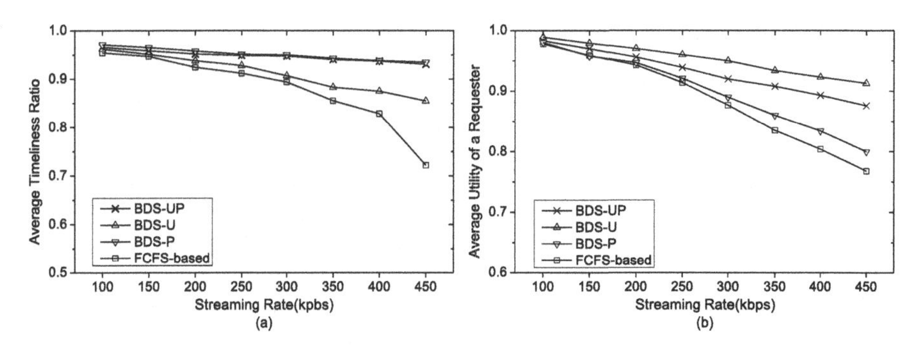

Fig. 5 (**a**) Timeliness ratio and (**b**) average utility after adding nodes with high bandwidth capacity

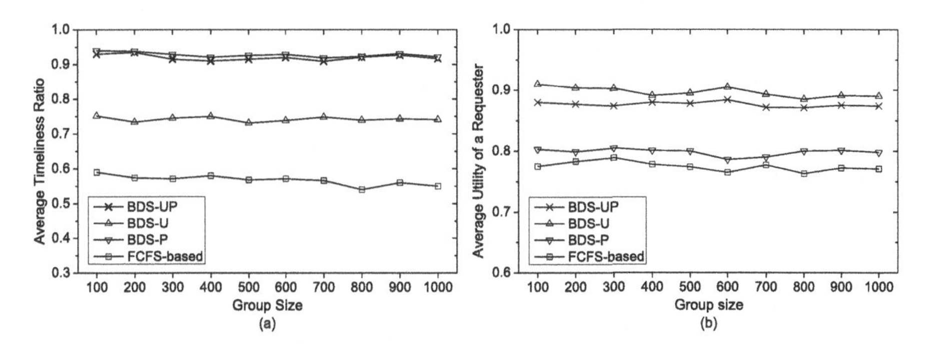

Fig. 6 (**a**) Timeliness ratio and (**b**) average utility for different group sizes

5 Conclusion and Future Work

In this paper, we analytically study the block delivery scheduling issue in the data-driven peer-to-peer streaming system and model the issue as a optimization problem mathematically. Moreover, we propose the optimal scheduling strategy based on the analytical model, which is helpful to improve the satisfaction degrees of nodes and the playback performance of the system. The effectiveness of our strategy is verified by extensive experiments. For future work, we will evaluate the performance of our strategy in the practical streaming system and improve the adaptivity of the strategy to the real network environment. We also would like to study how to design a more practical expression of the utility weight from the view of users in our strategy.

Acknowledgements. The work is supported by Guangdong Natural Science Foundation under grant No.S2011040006119.

References

1. Chou, C., Golubchik, L., Lui, J.C.S.: Design of scalable continuous media servers with dynamic replication. J. Multimedia Tools and Applications 17(2-3), 181–212 (2002)
2. Diot, C., Levine, B.N., Lyles, B., Kassem, H., Balensiefen, D.: Deployment issues for the IP multicast service and architecture. IEEE Network 14(1), 78–88 (2000)
3. Liu, Z., Wu, C., Li, B., Zhao, S.: UUSee: Large-scale operational on-Demand streaming with random network. In: Proceedings of IEEE INFOCOM (2010)
4. Huang, Y., Fu, T., Chiu, D., Lui, J.C.S., Huang, C.: Challenges, Design and Analysis of a Large-scale P2P-VoD System. In: Proceedings of ACM SIGCOMM (2008)
5. Zhang, X., Liu, J., Li, B., Yum, T.: Coolstreaming/donet: A data-driven overlay network for efficient live media streaming. In: Proceedings of IEEE INFOCOM (2005)
6. Hei, X., Liang, C., Liang, J., Liu, Y., Ross, K.W.: A measurement study of a large-scale P2P IPTV system. IEEE Transaction on Multimedia 9(8), 1672–1687 (2007)
7. Zhang, X., Liu, J., Li, B.: On large scale peer-to-peer live video distribution: Cool Streaming and its preliminary experimental results. In: Proceedings of IEEE International Workshop on Multimedia Signal Processing, MMSP (2005)
8. Pai, V., Kumar, K., Tammilmani, K., Sambamurthy, V., Mohr, A.E., Mohr, E.E.: Chainsaw: eliminating trees from overlay multicast. In: Proceedings of IEEE INFOCOM (2005)
9. Garbacki, P., Epema, D.H.J., Pouwelse, J.: Offloading servers with collaborative video on demand. In: Proceedings of the Seventh International Conference on Peer-to-Peer Systems, IPTPS (2008)
10. Mol, J.J.D., Pouwelse, J.A., Meulpolder, M., Epema, D.H.J., Sips, H.J.: Give-to-Get: free-riding resilient video-on-demand. In: Proceedings of Multimedia Computing and Networking, MMCN (2008)
11. Yang, Y., Chow, A., Golubchik, L., Bragg, D.: Improving QoS in Bit Torrent-like VoD systems. In: Proceedings of IEEE INFOCOM (2010)
12. Agarwal, V., Rejaie, R.: Adaptive multi-source streaming in heterogeneous peer-to-peer networks. In: Proceedings of Multimedia Computing and Networking, MMCN (2005)
13. Parvez, K.N., Williamson, C., Mahanti, A., Carlsson, N.: Analysis of Bit Torrent-like protocols for on-demand stored media streaming. In: Proceedings of ACM SIGMETRICS (2008)
14. Zhang, M., Xiong, Y., Zhang, Q., Sun, L., Yang, S.: Optimizing the throughput of data-driven peer-to-peer streaming. IEEE Transactions on Parallel and Distributed Systems 20(1), 97–110 (2009)
15. Tewari, S., Kleinrock, L.: Analytical model for Bittorrent-based live video streaming. In: Proceedings of Consumer Communications and Network Conference, CCNC (2007)
16. Bittorrent, `http://bitconjuer.com`
17. Bharambe, A.R., Herley, C., Padmanabhan, V.N.: Analyzing and improving a BitTorrent network's performance mechanisms. In: Proceedings of IEEE INFOCOM (2006)
18. Saroiu, S., Gummadi, P., Gribble, S.: A measurement study of peer-to-peer file sharing systems. In: Proceedings of Multimedia Computing and Networking, MMCN (2002)
19. Li, D., Cui, Y., Xu, K., Wu, J.: Segment-sending schedule in data-driven overlay network. In: Proceedings of IEEE ICC (2006)

IDES: Self-adaptive Software with Online Policy Evolution Extended from Rainbow

Xiaodong Gu

Abstract. One common approach or framework of self-adaptive software is to incorporate a control loop that monitoring, analyzing, deciding and executing over a target system using predefined rules and policies. Unfortunately, policies or utilities in such approaches and frameworks are statically and manually defined. The empirical adaptation policies and utility profiles cannot change with environment thus cannot make robust and assurance decisions. Various efficiency improvements have been introduced to online evolution of self-adaptive software itselfhowever, there is no framework with policy evolution in policy-based self-adaptive software such as Rainbow. Our approach, embodied in a system called IDES(Intelligent Decision System) uses reinforcement learning to provides an architecture based self-adaptive framework. We associate each policy with a preference value.During the running time the system automatically assesses system utilities and use reinforcement learning to update policy preference. We evaluate our approach and framework by an example system for bank dispatching. The experiment results reveal the intelligence and reactiveness of our approach and framework.

1 Introduction

As the complexity of the software systems increases and the environment becomes open and various, self-adaptive software is playing an increasingly important role in software systems. The primary self-adaptations are implemented by embedding some failure-handing code within the systems.However,this is not understandable and can't deal with complex and large amount of logics. Therefore, it is the most practicable way to separate business logics from self-adaptive system and defining external self-adaptive policies by users or experts. General approach for adaptation today is to incorporate a feedback control loop to adjust target system through

Xiaodong Gu
Nanjing University, Nanjing, China
e-mail: guxiaodong1987@126.com

R. Lee (Ed.): Computer and Information Science 2012, SCI 429, pp. 181–195.
springerlink.com

monitoring its context, detecting significant changes, deciding how to react, and acting to execute such decisions with the addition of a shared knowledge-base[1].The knowledge-base here mainly includes self-adaptive policies such as rules or strategies implying how and what to do under self-adaptive demands.For the system adaptation, adaptive systems usually take real-time assessment called *system utility* for target system.Meanwhile,the adaptive operations to the system will bring a certain feedback called *cost-benefit*.

There are varieties of researches and practice about the policy-based self-adaptation theories[1, 2, 3, 4]. Rainbow[3],presented by Garlan, et al is one of the most outstanding self-adaptive software that uses architecture-based techniques combined with control and utility theories. It not only provides a general, supporting mechanism for self-adaptation of different classes of systems but defines a language, called *Stitch*, that allows adaptation expertise to be specified and reasoned about [10]. Since the Rainbow system has made a great success, researchers have concerned much of its improvements. For example, Rahul Raheja et al. gave a number of improvements such as conflict resolution of strategies and concurrently by increasing the dimension of time delay [4]. Most of all, as Garlan himself pointed in his later research[10], such adaptive software suffers the shortcoming of:

- ***Policy and Utility Stereotyped:***
  The utility preference profiles and cost-benefit attributes in the Rainbow's *stitch* language are statically determined that can't change with external environment once assumed, thus make the adaptation software unreactive when facing unexpected exceptions.
- ***Frail Decisions and Low Reaction Efficiency:***
  When in an exceptional case, the self-adaptive system cannot react timely. Thus gives frail and non robust decisions.

He then introduced a concept named *adapting adaptation* which looked into deep researches about online policy evolution for self-adaptation software itself.

To cope with the problems discussed above, FUSION[5] adds another control loop that enables evolution of self-adaptive logics. However, since the adaptation and evolution are based on system features which are domain dependent and not easy for definitions, additions and deletions, FUSION's evolution is not loose coupling with adaptation domains in comparison to policy-based self-adaptive software such as Rainbow[17]. Kim et. al presented concepts of off-line and on-line planning for self-management software[6] that focus on self-management at runtime instead of design state. They use reinforcement learning for adaptation, but their adaptation methods are just for the target systems themselves, there are no corresponding solutions and application examples for policy-based adaptive software evolution. Two more reasons can motivate the development of an extension of Rainbow for policy evolution:

- The policy-based self-adaptive software such as Rainbow is domain-independent and is very good at policy definition, addition and deletion.
- There have not been any online evolution mechanism such as double control loop for Rainbow, that is, the adapting adaptation problem in Rainbow is not solved.

This paper presents a framework called IDES that extends the Rainbow and its *stitch* language with policy evolution mechanism. First and foremost,we give some definitions about online evolution of self-adaptive software.

Definition 1. (Online Evolution of Self-adaptation)The online evolution of self-adaptation evolution,also called self-adaptive software evolution, means the self-adaptation logics or adaptation decisions itself can change with external environment in self-adaptive software.

Definition 2. (Online Policy Evolution of Self-adaptation)The online policy evolution of self-adaptation evolution means the self-adaptation policies or strategies defined for self-adaptation can change with external environment particularly in the policy based self-adaptive software.

In the framework,we introduce a reinforcement learning based policy evolution method. Instead of statically defining the adaptive policies and empirically defining the cost-benefit attributes at develop time, we provide online reassessments and evolutions of policy preference, cost-benefit attributes, etc. Thus give more intelligent and flexible decisions to target systems.

We elaborate on two key contributions of IDES :

- It gives a model free online evolution method for policy based self-adaptive software systems.
- Some AI techniques are introduced such as the intelligent decision and the reinforcement learning to make itself-adaptation more intelligent and responsive.

The remainder of the paper is organized as follows. Section 2 motivates the problem with an example system that also serves as a case study in this paper. Section 3 presents our improvements in Rainbow. Section 4 provides an overview of IDES. Respectively detail IDES's policy-based model of adaptation and its learning mechanism. Sections 5 presents the case study and evaluation details of IDES. Section 6 briefly introduces some related work. Section 7 concludes with an overview of the related work and future avenues of research.

2 Background and Motivation

2.1 The Original Rainbow Self-adaptive Software

The Rainbow framework uses the software architecture and a reusable infrastructure to support self-adaptation of software systems[4]. Figure 1 illustrates its adaptation control loop. Probes extract information from the target system. Gauges get the information and abstract them into architecturally relevant information. The architecture evaluator checks constraints in the model and triggers adaptation if any violation is found. The adaptation manager, on receiving the adaptation trigger, chooses the best strategy to execute, and passes it to the strategy executor, which executes the strategy on the target system via effectors.

Fig. 1 Rainbow framework [3]

2.2 A Motivational Example

We illustrate the motivation of policy evolution by an example of an Online Bank Dispatch System (BDS) and its self-adaptation with Rainbow. The BDS is a simulation system about a service scenario in a bank including customers and cashiers. There are some cashiers each of whose service time is under the normal distribution with mean = 8, 10, 12 min. If a customer found there are 6 or more people waiting in line when arriving, he will leave with a probability of 0.3 while the rest continue waiting. In order to improve the quality of service, the scheduling system need to adjust the cashier allocation such as increasing the cashier or choose a more skilled cashier when the queue is long enough. Figure2 shows its software architecture in the traditional component-and-connector view. BDS aims to allocating cashiers to minimize the average queue length thus gives the quickest service. The system is required to attain a number of QoS goals, such as a short queue length, less quantity of left persons.BDS needs to be self-adaptive to deal with QoS dissatisfactions, such as too long queue length, too many persons left. For instance, adding cashiers or replacing a cashier with a more skill one.

In Rainbow, we can define a number of strategies for this adaptation, for example:

- ***Strategy s1:*** if queue length>6 then replace a cashier with a more skill one. With a cost-benefit of 10
- ***Strategy s2:*** if queue length>6 then add a cashier. With a cost-benefit of 5

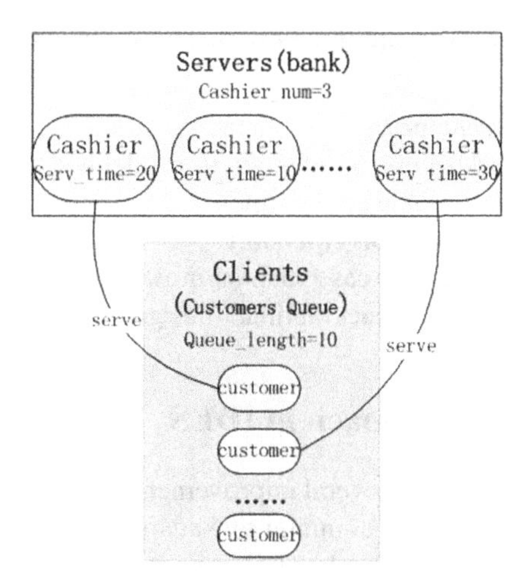

Fig. 2 Architecture of BDS

For strategy selection, we can assess each strategy by its long term utility in each quality dimension. We firstly define utility profiles and preferences for each quality dimension, as table 1 shows. Then, we define the cost-benefit attribute vectors for each strategy, as table 2 showed.

Table 1 BDS utilities and preference in stitch language

quality dimension	description	utility	preference
U_{ql}	queue length	(short,1),(medium,0.5),(long,0)	0.8
U_{aw}	average waiting time	(short,1),(medium,0.5),(long,0)	0.2

Table 2 BDS cost-benefit attributes vectors and utility evaluation for applicable strategies in stitch language

strategy	U_{ql}	U_{aw}	strategy	weighted utility evaluation
S1	+2	+0.5	S1	U=0.8*2+0.2*1=0.18
S2	+1	+0.01	S2	U=0.8*0.5+0.2*0.01=0.42

During running, Rainbow adaptation engine assesses the utility evaluation by calculating utility evaluation for each strategy, as table 2 shows. Then, the strategy S1 with the highest evaluation 0.18 will always be selected.

As mentioned earlier, there are three problems associated with the construction of adaptation logic:

- ***Policy and utility stereotyped:***
  The policies and cost-benefit of adaptation are defined before running, so the adaptive decision is invariable all the time.
- ***Frail decision and low reaction efficiency:***
  When there is an exceptional case such as most of cashiers go for holiday, the self-adaptive system cannot react on time. thus give frail decisions.

3 Policy Evolution Approach in IDES

In this section we will elaborate several improvements of Rainbow and its *stitch* language. First of all, we model the common self-adaptation process in Rainbow with MDP model[8], then work out some key issues on improving self-adaptive language before evolution of adaptation policies. Finally we introduce a reinforcement learning based method for policy evolution online.

3.1 Markov Process Modeling of Policy-Based Adaptation

Definition 3. (markov decision process) A Markov Decision Process (MDP) is a quadruple ⟨S,A,R,P⟩,where S is a state set, A is an action set, R:SA→R is a reward function. P:SA→PD(S) is a state transfer function. P(s,a,s') denotes the transfer probability from state s to s' via action a. The essence of MDP is that the probability of next state transfer is depended only on the current state and action, rather than historical states.

To formally use our reinforcement learning based method, we bring the MDP model in the self-adaptive processes, S denotes a set of system states. A denotes a set of adaptation policies. State transfer matrix P denotes the state transfer after an adaptation process. R denotes the optimizing extent after an adaptation process.

3.2 Strategy Selection and Conflict Resolution

Strategy selection occurs in the common situation that more than one strategy could be used for decision at one time,we should choose the best one that could lead to the highest utility [11, 12]in the long run. For each strategy, we define a *preference value* which represents the preference of selecting that strategy when there is more than one strategy for current adaptation. The higher preference value, the more probability of the strategy to be selected. We call this process *strategy selection* or *conflict resolution* which give us an appropriate adaptation interface for online evolution of adaptation logics. In *stitch*, strategy preferences are calculated by utilities and cost-benefit values that are predefined empirically and stationary while our evolution goal discussed in section 3.4 is to adapt the preference values for every strategies by learning so as to support the online evolution of self-adaptation. Equation 1 shows the calculation of selecting a strategy

$$\pi_t(s,t) = Pr(a_t = a|s_t = s) = \frac{e^{p(s,a)}}{\sum_{b \subseteq S'} e^{p(s,b)}} \tag{1}$$

where p(a) is the preference value for decision action a and S' is a set of decisions matching current condition.

3.3 Online Cost-Benefit Assessments

The premise of optimizing the decision is the evaluation we took on the target system after executing the decision.We denote this evaluation as *cost-benefit*[3]. Typically in Rainbow, the cost-benefit is defined by users manually. However, in real systems it is difficult for users to estimate the utility of a strategy beforehand. Besides, the environment is changing all the time leading to the variability of utility. An ideal solution of this estimation is to calculate the real utility based on the real time condition. As a result,we use the variation of system utility to define the cost-benefit value. Assume that the system utility during previous decision is U_{t-1}, and is U_t currently, then the cost-benefit value respected to previous decision is

$$R_{t+1} = \begin{cases} 1, & \text{for } U_t > U_{t-1}; \quad (2) \\ -1, & \text{for } U_t < U_{t-1}. \quad (3) \end{cases}$$

3.4 Policy Evolution Method

To evolve the adaptive policies based on system utility online, we introduce a policy evolution method based on reinforcement learning [8].Reinforcement Learning (RL) provides a promising new approach to systems performance management that differs radically from common approaches making use of explicit system performance models. In principle, RL can automatically learn high-quality management policies without an explicit performance model or traffic model, and with little or no built-in system specific knowledge [7]. We use a RL algorithm named *actor-critic*[9]. It assumes that there is an actor who takes an action by a policy. A critic evaluates the performance by the reward from environment and optimizes the policy through updating the value functions. Our goal is to make a model correspondence between them and bring the actor-critic algorithm in our evolution method. Table 3 illustrates this correspondence. After corresponding, the evolution method can concluded as algorithm 1.

4 IDES Framework

Figure 3 illustrates the IDES framework. Like Rainbow, it is architecture based self-adaptive software, but extending from it, strategies, utility profiles and cost-benefit

Table 3 correspondences between actor-critic algorithm and self-adaptive software evolution method

actor-critic algorithm	**Self-adaptive software evolution method**
state S	system condition
State-action value Q(s,a)	The preference value of a strategy in current condition P(condition,strategy)
Policy (S,a)	The probability of selecting strategy S in condition c $$Pr(srategy = s \mid condition = t) = \frac{e^{P(s,a)}}{\sum_{b \subseteq S'} e^{P(s,b)}} \quad (4)$$
Real time reward r_t	Cost-benefit of adaptation decision $Rt = U_t - U_{t-1}$

Algorithm 1. Policy optimizing method for self-adaptive software.

1: initiate preference value of each strategy as user defined;
2: select a strategy strategy0 according to the current system state;
3: **while** true **do**
4: execute strategy0observe the effect and get the reward vector $R_t = U_t - U_{t-1}$;
5: analyze system ,get a sub set S of strategies that can be used for current adaptation decision;
6: update preference of current strategy strategy0 based on sub strategy set

$$\delta_t \leftarrow R_t + \gamma \max_{s \subseteq S} P(s) - P(strategy0) \quad (5)$$

$$P(strategy0) \leftarrow P(strategy0) + \alpha \gamma_t \quad (6)$$

7: select the next strategy from sub strategy set S according to the probability calculated by preference value

$$strategy0 = \arg\max_{s \subseteq S} \frac{e^{P(s,a)}}{\sum_{b \subseteq S'} e^{P(s,b)}} \quad (7)$$

8: **end while**

attributes in the IDES framework can be evolved online by updating the preference of each policy during running time. It is implemented by two close control-loops. One loop monitors the running system, analyzes the adaptive demand and experts appropriate decision, another loop evaluates system utility and updates strategy preference using machine learning. Besides, it has an adaptation knowledge management module for defining, interpreting and managing the adaptation knowledge. The rest of this section further elaborates on the knowledge management model and two control-loops.

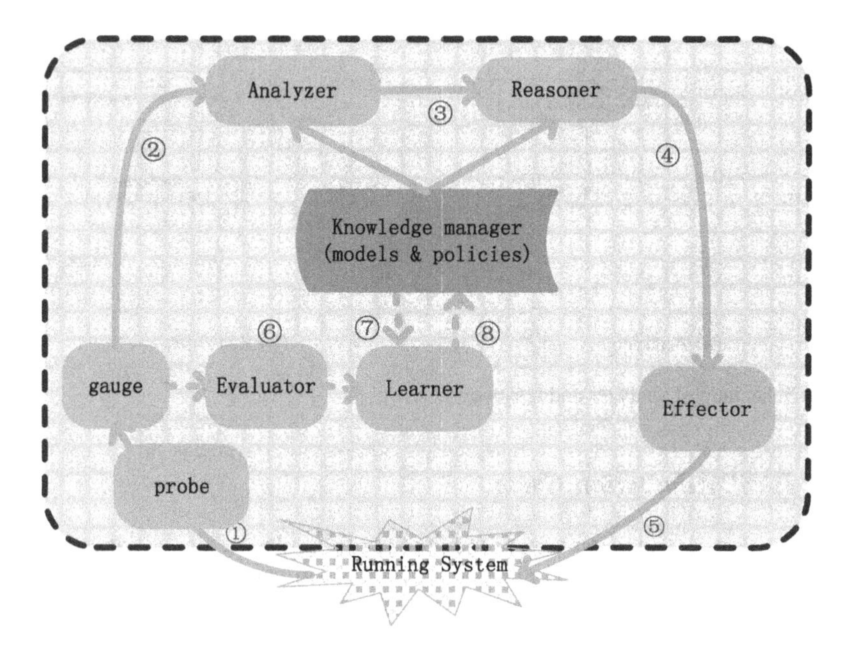

Fig. 3 IDES framework

4.1 IDES Adaptation Knowledge Management

As architecture based self-adaptive software, we firstly need some descriptions of the target system's architecture and some adaptation expertise. Therefore, we design an adaptive language to represent the adaptation knowledge using high-level adaptation concepts. In Rainbow, the self-adaptive language is *stitch* which consists of *operators*, *tactics*,*strategies*,*quality dimensions*, *utility preferences* and so on[3]. As for supporting of policy evolution, we extend some elements from the *stitch* language and developed a self-adaptive language called *"SASGDL"*(a General Descriptive Language for Self-Adaptive Systems) which consists of *family*, *instance*, and *system utility* etc. We also make some descriptions of system operators and adaptive strategies. Figure 4 is a highlight of some grammars of the adaptation description language.

Table 4 adaptation description language grammar highlights

System	::=	family|instance|strategy|tactic|operator |probe|gauge|systemutility
Family	::=	properties|component|connector
Instance	::=	properties|component|connector
Strategy	::=	condition,referencePolicy
Tactic	::=	condition,operator,effect
Operator	::=	type,arguments,returnvalue

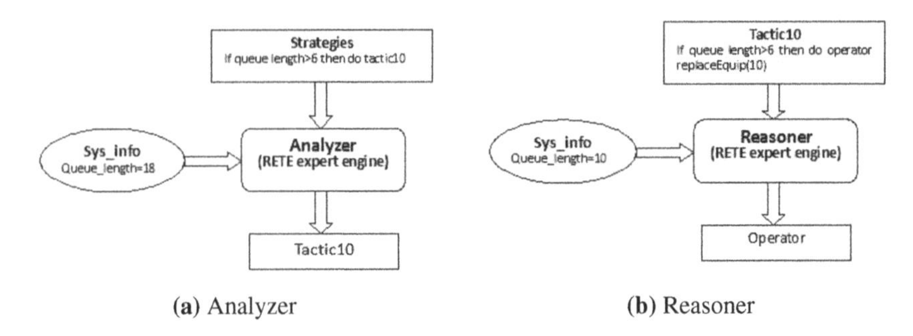

(**a**) Analyzer (**b**) Reasoner

Fig. 4 Schematic of analyzer and reasoner

4.2 IDES Adaptation Cycle

In this section, we describe how to monitor, analyze, reason, and execute with the concept model and adaptation policies to adapt a software system in IDES. As is showed in figure 3, IDES consists of four parts which form the adaptation cycle. They are *probe*, *gauge*, *analyzer*, *reasoner* and *effector*.

Probe and gauge. The *probe* and *gauge* act as monitor in our self-adaptive system. The *probe* get variables of target system and the *gauge* calculate the properties in architecture view. They provide system information for reasoning engine for further decisions.

Analyzer. As is illustrated in figure 4a, The *analyzer* gets the system information from the *gauge* and analyzes latent adaptive demand by an expert engine which runs the RETE algorithm [15].

Reasoner. As is illustrated in figure 4b, After analyzing, the *reasoner* gets the adaptation requirements (tactics) derived from analyzer and also gets system information from gauge. Then it reasons out concrete operators by an expert engine which runs the RETE algorithm.

Executer. The executer gets the operators derived from the reasoner and executes them through system operation module.

4.3 IDES Policy Evolution Cycle

In this section, we describe how to evaluate system utility and evolve policies by our method in IDES. As is showed in figure 3, the evolution cycle include *evaluator* and *updater*.

Evaluator. The *evaluator* get system information from *gauge* and calculate system utility by the system utility file that user have defined. It gives the utility value to updater for updating the current policies.

Updater. The *updater* gets system utility value from *evaluator*, and then executes the online policy evolution method discussed above to update the policies' preference by historical policies and current system utility value.

5 Case Study

In order to adapt the bank scheduling system exemplified in section 2,we made a simulation of target system - BDS and incorporate the IDES into it. During running,IDES monitors the BDS system and fetches some running information such as queue length, analyzing adaptation demands and adapting them with a strategy when the queue length is over a certain threshold. We use the same two strategies (strategy1 and strategy2) as exemplified in section 2. To compare the policy evolution technique with other original self-adaptation,we take experiments in three conditions:

- ***Non Adaptive:*** To run the BDS without IDES.
- ***Adaptive without Policy Evolution:*** Run the BDS with IDES which doesn't have the updater modules,that is to say, the IDES's functions are similar to Rainbow that doesn't have policy evolution mechanisms.
- ***Adaptive with Policy Evolution:*** Run the BDS with IDES which has the updater modules,namely has the online policy evolution mechanism.

We should evaluate three aspects: 1. adaptation performance; 2. evolution performance include self intelligent and effect on global adaptation; 3 evolution effectiveness.

5.1 Adaptation Performance

The adaptation performance means the function of adaptation with target system. To evaluate the basic self-adaptation capability, we executed the scenario mentioned above both without any self-adaptation and with self-adaptation. Figure 5 shows the queue length difference between both cases. Without adaptation, the queue length will be persistently increasing while after adaptation it is kept fewer than 6 all the while. This proves a prominent adaptation efficient of our IDES self-adaptive system.

5.2 Intelligence and Evolution Effect

We define intelligence here as that the adaptive software can quickly reassessment the pre-defined policies in case of unexpected or unusual changes in the environment thus make better decisions. To evaluate this ability, we add an exception during adaptation that there are only 3 cashiers available, so strategy1 will make no sense in this case. Figure6 shows the variation of two strategies' preference value during the experiment. As we can see in figure6a, without policy evolution, the self-adaptive software is helpless when faced with the exception.As is indicated by the green line,

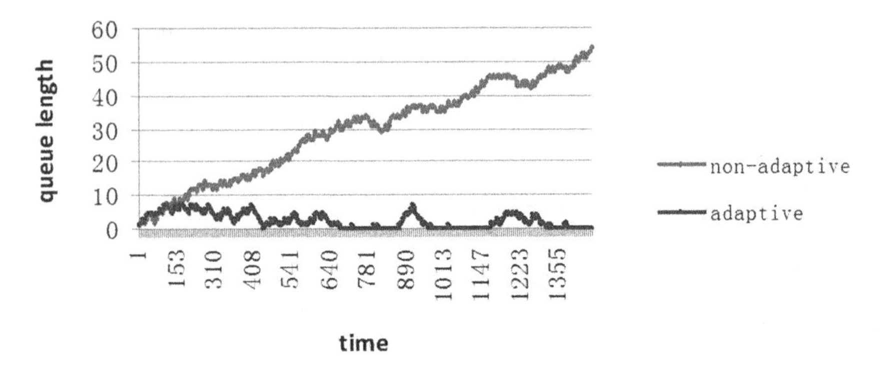

Fig. 5 Result comparison between non adaptive and adaptive

the queue length of BDS keeps growing without policy evolution method,however, when add online policy evolution, as figure6b shows, the queue length is drop down to normal quickly after a short time of growth, which indicates that the policies can change their preferences quickly when having detected exception and a better strategy(strategy2) will be chosen , thus gives a more appropriate and robust adaptation and makes the BDS service more stable and reliable. This experiment shows that while using our approach, the self-adaptive software is more intelligent. And besides, it convicts effects of the online policy evolution mechanism when facing abnormalities.

6 Related Work

Over the past decade, researchers have developed a variety of methodologies, technologies, and frameworks to construct self-adaptive systems [1][3]. We provide an overview of the state-of-art in this area and compare them with IDES.

6.1 Architecture Based Self-adaptive Software

Architecture based self-adaptive software is to model the target system in an architecture view and to give self-adaptive during running time. An architecture model provides a global perspective on the system and exposes the important system-level behaviors and properties, thereby helping to the run time monitoring and adaptive decisions [10]. The typical framework is Rainbow [3, 10], which not only provides a general, supporting mechanisms for self-adaptation over different classes of systems but defines a language called *Stitch*, that allows adaptation expertise to be specified and reasoned about. However, the *Stitch* language has a drawback that both the policies and cost-benefit properties are manually predefined and cannot update online. Our IDES adaptive system is mainly to address this problem. We realized the basic

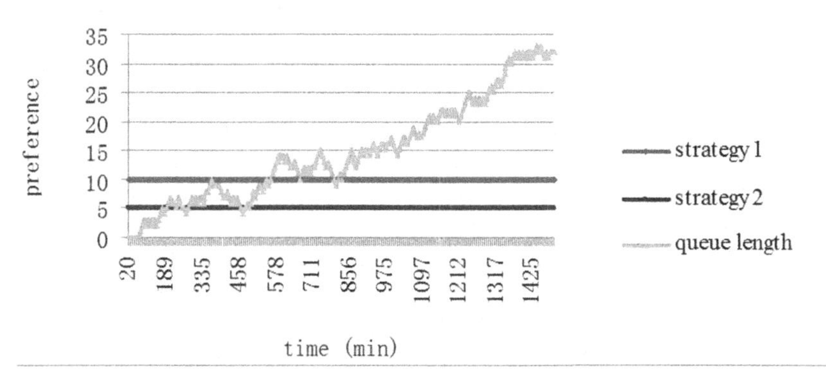

(**a**) queue length under exception(cashier insufficient) but without online policy optimizing

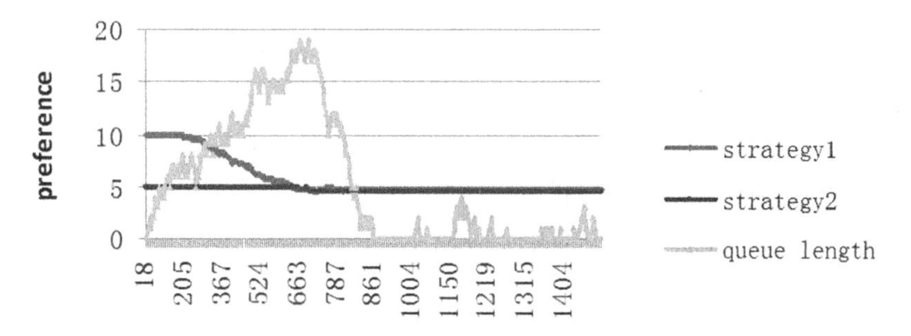

(**b**) Queue length under exception (cashier insufficient) with online policy optimizing

Fig. 6 Preference value and queue length of two strategies during the experiment

functional module of Rainbow and besides, we added the online policy evolution mechanism for the evolution of self-adaptive system itself.

6.2 Online Evolving Self-adaptive Software

The online evolution of self-adaptive software has already been gained attentions in the past years. Elkhodary et al. firstly introduced a double control-loop for self-adaptation, one for adaptation and the other for self learning [5]. However, as discussed in section 1, their adaptation systems are feature based which is not easy for definition, additions and deletions. Besides, its evolution method is based on feature selection which is limited by feature space. The IDES we present can be viewed as a practical improvement which not only has learning loop but has policy based intelligent decisions where policies are easy for definition,addition and deletion.

6.3 Reinforcement Learning Based Self-adaptive Software

Kim et al. firstly used reinforcement learning techniques for online self adaptation on general self-management systems [6]. They pointed out the difference between online update and offline update and verified the advantages of online update, however, their optimizing was just for the target system itself, there are no corresponding solutions and application examples for policy-based adaptive software evolution while reinforcement learning in our method is focused on policy evolution in policy-based adaptive software.

Table 5 gives a comparison of some typical self-adaptive software. From a view of adaptive target, IDES has a relatively general usage scope. As for adaptive approach, IDES is policy based that is more flexible for policy addition and deletion. Most of all, It has the outstanding features of RL-based and online self-adaptation, which make it more intelligent and reactive with the change of environment.

Table 5 the comparison of several self-adaptive software

	adaptive target	**adaptive approach**	**self-optimizing ability**	**online or offline**
MAPE	Architecture	policy-based	disable	offline
Rainbow	Architecture	policy-based	disable	offline
FUSION	Architecture	feature-based	able	online
Kim et al.2009	Robot	RL-based	-	online
IDES	System	policy-based	RL-based	online

7 Conclusion

Rainbow,as a self-adaptive software, provides reliable and stable services through the monitoring, analyzing, decision-making and execution over target system in variable environment but is not flexible with operating environment mutation or system anomalies because their adaptive policies are statically specified at the design stage. We extend the Rainbow with a framework named IDES in which an on-line policy evolution method based on reinforcement learning is presented to solve problems about strategic choices, conflict resolution, cost-effective reassessment and online policy evolution. A typical application scenario that we realized with IDES verifies the feasibility and effectiveness of our approach. It also shows the advantages of online evolution in policy-based self-adaptation.

Our future work includes the evolution on specific policy contents. Furthermore, more AI techniques such as generic algorithms can be used to serve this purpose.

References

1. Salehie, M., Tahvildari, L.: Self-adaptive software: Landscape and research challenges. ACM Transactions on Autonomous and Adaptive System 4(2), 1–42 (2009)
2. Kephart, J.O., Chess, D.M.: The Vision of Autonomic Computing. IEEE Computer 36(1), 41–50 (2003)

3. Cheng, S.-W.: Rainbow: Cost-Effective Software Architecture-Based Self-Adaptation. PhD thesis, Carnegie Mellon University, Pittsburgh, PA. Technical Report CMU-ISR-08-113 (2008)
4. Raheja, R., Cheng, S.-W., Garlan, D., Schmerl, B.: Improving Architecture-Based Self-adaptation Using Preemption. In: Weyns, D., Malek, S., de Lemos, R., Andersson, J. (eds.) SOAR 2009. LNCS, vol. 6090, pp. 21–37. Springer, Heidelberg (2010)
5. Elkhodary, A., Esfahani, N., Malek, S.: FUSION: A Framework for Engineering Self-Tuning Self-Adaptive Software Systems. In: The Eighteenth ACM SIGSOFT International Symposium on Foundations of Software Engineering, pp. 7–16 (2010)
6. Kim, D., Park, S.: Reinforcement learning-based dynamic adaptation planning method for architecture-based self-managed software. Workshop on Softw. Eng. For Adaptive and Self-Managing Systems, 76–85 (2009)
7. Tesauro, G., et al.: A Hybrid Reinforcement Learning Approach to Autonomic Resource Allocation. In: Int'l Conf.on Autonomic Computing, pp. 65–73 (2006)
8. Barto, A.G., Sutton, R.S., Watkins, C.J.C.H.: Learning and sequential decision making. Technical Report 89-95, Department of Computer and Information Science. University of Massachusetts, Amherst, Massachusetts (1989)
9. Konda, V.R.: Actor-Critic Algorithms. Ph.D. thesis, Department of Electrical Engineering and Computer Science, Massachusetts Institute of Technology (2002)
10. Garlan, D., Schmerl, B., Cheng, S.-W.: Software Architecture-Based Self-Adaptation. Autonomic Computing and Networking, 31–56 (2009)
11. Walsh, W.E., Tesauro, G., Kephart, J.O., Das, R.: Utility functions in autonomic systems. In: International Conference on Autonomic Computing (2004)
12. Alia, M., Eide, V.S.W., Paspallis, N., Eliassen, F., Hallsteinsen, S., Papadopoulos, G.A.: A Utility-based Adaptivity Model for Mobile Applications. In: Proceedings of the IEEE International Symposium on Ubisafe Computing (UbiSafe 2007). IEEE Computer Society Press (2007)
13. Kakousis, K., Paspallis, N., Papadopoulos, G.A.: Optimizing the Utility Function-Based Self-adaptive Behavior of Context-Aware Systems Using User Feedback. In: Proceedings of the OTM 2008 Confederated International Conferences, Coopis, Doa, Gada, Is, and ODBASE (2008)
14. Abdelwahed, S., Kandasamy, N., Neema, S.: A control-based framework for self-managing distributed computing systems. In: Proceedings of the 1st ACM SIGSOFT Workshop on Self-Managed Systems, Newport Beach, California, pp. 3–7 (2004)
15. Sottara, D., Mello, P., Proctor, M.: A Configurable Rete-OO Engine for Reasoning with Different Types of Imperfect Information. IEEE Transactions on Knowledge and Data Engineering 22(11), 1535–1548 (2010)
16. Cheng, B.H.C., de Lemos, R., Giese, H., Inverardi, P., Magee, J., Andersson, J., Becker, B., Bencomo, N., Brun, Y., Cukic, B., Di Marzo Serugendo, G., Dustdar, S., Finkelstein, A., Gacek, C., Geihs, K., Grassi, V., Karsai, G., Kienle, H.M., Kramer, J., Litoiu, M., Malek, S., Mirandola, R., Müller, H.A., Park, S., Shaw, M., Tichy, M., Tivoli, M., Weyns, D., Whittle, J.: Software Engineering for Self-Adaptive Systems: A Research Roadmap. In: Cheng, B.H.C., de Lemos, R., Giese, H., Inverardi, P., Magee, J., et al. (eds.) Software Engineering for Self-Adaptive Systems. LNCS, vol. 5525, pp. 1–26. Springer, Heidelberg (2009)
17. Anthony, R.J.: A Policy-Definition Language and Prototype Implementation Library for Policy-based Autonomic Systems. In: Proceedings of IEEE International Conference on Autonomic Computing, pp. 265–276 (2006)

Visualization of Logical Structure in Mathematical Proofs for Learners

Takayuki Watabe and Yoshinori Miyazaki

Abstract. This study focuses on the visual representation of mathematical proofs for facilitating learners' understanding. Proofs are represented by a system of sequent calculus. In this paper, the authors discuss SequentML, an originally designed XML (Extensible Markup Language) vocabulary for the description of sequent calculus, and the visualization of mathematical proofs by using this vocabulary.

Keywords: E-learning, mathematical proof, visualization, sequent calculus, MathML.

1 Introduction

This paper discusses a framework for visualizing mathematical proofs in order for learners to better understand these proofs. In the field of math education, there has been some previous research on the methodology of teaching how to solve proofs [1, 2], and our study primarily focuses on the logical structure of proofs. Proofs are usually written in natural languages, and this sometimes causes learners to face difficulty in understanding the logical structure of these proofs. The authors believe that learners do not comprehend the proofs or acquire the ability to describe proofs because they do not understand the logical structure of these proofs. In fact, [3] shows that the more structured proofs were presented to learners, the higher their scores were, in the course in propositional and predicate logic. Although formal languages help us to show rigidly the logical structure of proofs, grasping how to read these languages is not easy. Therefore, this study aims at the visualization of the logical structure of mathematical proofs.

Takayuki Watabe
Graduate School of Informatics, Shizuoka University, Shizuoka, 432-8011, Japan
e-mail: gs11055@s.inf.shizuoka.ac.jp

Yoshinori Miyazaki
Faculty of Informatics, Shizuoka University, Shizuoka, 432-8011, Japan
e-mail: yoshi@inf.shizuoka.ac.jp

R. Lee (Ed.): Computer and Information Science 2012, SCI 429, pp. 197–208.
springerlink.com

It is desirable to use the expressions of proofs that show their logical structure in a rigid and clear manner, and have a high compatibility with visualization. In order to realize this, a system of sequent calculus is adopted.

Sequent calculus expresses the proposition to prove by a style called sequent, and expresses proofs by showing the process in which the sequent is transformed into a relatively simple form. Because the method to express the transformation process of sequents as a type of diagram is already known, visualization is attempted so that even novice learners easily understand mathematical proofs. As one of the related studies, Smullyan [4] expressed proofs by using logical diagrams.

For describing sequent calculus, we propose an originally designed XML (Extensible Markup Language) vocabulary called SequentML. MathML (Mathematical Markup Language), an XML vocabulary recommended by W3C (World Wide Web Consortium) as one of the XML vocabularies for expressing math, is also used, because proofs involve math expressions. A high usability between MathML and SequentML are expected because both are XML vocabularies.

In section 2, MathML is introduced as one of the XML vocabularies for describing math expressions. Section 3 outlines the sequent calculus upon which representation of proofs is based in this research, and elaborates on the proposed SequentML to represent sequent calculus as XML data. In section 4, our method for visualizing proofs is presented with an example. Section 5 discusses the advanced functions of the proposed method. Section 6 presents the concluding remarks and plans for future work pertaining to this study.

2 MathML

MathML [5] is one of the XML vocabularies used for expressing math and has been released by W3C. There are two MathML syntaxes, which are Presentation markup and Content markup. The former affords complete control over the appearance of expressions, whereas the latter preserves the semantics of expressions. In this study, we use Content markup and we refer to it as MathML. Among the various tags defined in MathML, the <apply> tag, which represents application of a function or an operator, is central to MathML. Fig. 1 shows the MathML code for the expression $a + b$, with the tags explained in Table 1.

```
<math>
    <apply>
        <plus/>
        <ci>a</ci>
        <ci>b</ci>
    </apply>
</math>
```

Fig. 1 Representation of $a + b$ using MathML

Table 1 Description of tags in Fig. 1

Tag	Description
math	encloses MathML elements
apply	applies the first child operator or function to other elements
plus	performs addition
ci	encloses identifiers (variables and names)

Specifically, the code in Fig. 1 is interpreted as "applying addition to identifiers a and b." This notation also allows us to describe symbols used in set theory and logical operators such as

- "∈" (in), "⊂" (subset), "∩" (intersection), "∪" (union)
- "∧" (and), "∨" (or), "¬" (not), "⇒" (implies), "⇔" (equivalent), "∀" (for all), "∃" (exists).

To use these symbols or operators, one has to replace only the <plus> tag in Fig. 1 with the corresponding tag from one of abovementioned tags. Likewise, one may describe differentiation, integration, or matrices (introductions omitted owing to space constraints).

3 Sequent Calculus & SequentML

This section outlines a system of sequent calculus [6] upon which representation of proofs is based in this research, and elaborates on the proposed SequentML, an originally designed XML vocabulary to represent sequent calculus as XML data.

3.1 Sequent Calculus

In sequent calculus, the propositions to prove are presented in a style called sequent. A sequent is of the form in Fig. 2, where $\mathfrak{A}$, $\mathfrak{B}$, $\mathfrak{C}$, and $\mathfrak{D}$ denote logical expressions. This sequent is intuitively interpreted as "if $\mathfrak{A}$ and $\mathfrak{B}$ then $\mathfrak{C}$ or $\mathfrak{D}$ (holds)." In other words, the arrow symbol and its left and right commas represent "IMP (logical implication)," "And (logical product)," and "Or (logical disjunction)," respectively. The form "$\rightarrow \mathfrak{E}$" is also acceptable, and it simply means "to be $\mathfrak{E}$."

The transformation of sequents, or inference, is represented as in Fig. 3. This shows "$\mathfrak{A} \rightarrow \mathfrak{B}$; therefore, $\mathfrak{C} \rightarrow \mathfrak{D}$ (holds)." It may be retold that the sequent above a line ("$\mathfrak{A} \rightarrow \mathfrak{B}$") is simpler than that below ("$\mathfrak{C} \rightarrow \mathfrak{D}$"). There is also a case where there are multiple sequents above a line (Fig. 4). This implies that "$\mathfrak{E} \rightarrow \mathfrak{F}$" is proved if both "$\mathfrak{A} \rightarrow \mathfrak{B}$" and "$\mathfrak{C} \rightarrow \mathfrak{D}$" are shown. In this study, we consider the set theory for the examples given later.

$$\mathfrak{A}, \mathfrak{B} \rightarrow \mathfrak{C}, \mathfrak{D}$$

Fig. 2 Example of sequent

$$\frac{\mathfrak{A} \rightarrow \mathfrak{B}}{\mathfrak{C} \rightarrow \mathfrak{D}}$$

Fig. 3 Example of inference

$$\frac{\mathfrak{A} \rightarrow \mathfrak{B} \qquad \mathfrak{C} \rightarrow \mathfrak{D}}{\mathfrak{E} \rightarrow \mathfrak{F}}$$

Fig. 4 Example of multiple sequents above a line

3.2 SequentML

The authors developed an originally designed XML vocabulary to represent sequents. It is referred to as SequentML. SequentML enables us to write sequents and their transformations (or inferences). Table 2 presents a list of tags used for representing sequents.

Table 2 List of tags used for sequent representation

Tag	Description
sequent	encloses sequent
antecedent	left proposition of "→"
succedent	right proposition of "→"

Each proposition is expressed using MathML. Fig. 5 shows the SequentML codes for the sequent "$A \subset B, x \in A \rightarrow x \in B$."

```
<sequent>
    <antecedent>
        <math>
            <apply>
                <subset/>
                <ci>A</ci>
                <ci>B</ci>
            </apply>
        </math>
        <math>
            <apply>
                <in/>
                <ci>x</ci>
                <ci>A</ci>
            </apply>
        </math>
    </antecedent>
    <succedent>
        <math>
            <apply>
                <in/>
                <ci>x</ci>
                <ci>B</ci>
            </apply>
        </math>
    </succedent>
</sequent>
```

Fig. 5 "$A \subset B, x \in A \rightarrow x \in B$" by SequentML

A sequent is enclosed by the <sequent> tag. The logical expressions on the left of the arrow (antecedent logical expressions) are enclosed by the <antecedent> tag. Each logical expression is enclosed by the <math> tag, because they are written in MathML. When there are multiple logical expressions in the antecedent, each logical expression is described as a child of the <antecedent> tag. The logical expressions on the right of the arrow (succedent logical expressions) are also described in the same manner using <succedent> tag.

Next, the tags for describing a proof as a series of sequents are explained. These tags are shown in Table 3.

Table 3 List of tags used for representing inferences

Tag	Description
sequentcalculus	encloses sequent calculus
deduction	describes a single inference
rule	describes an inference rule

Let S1 and S2 be sequents. Assume that the inference from S1 to S2 is made on the basis of the definition of subsets. Further, assume that S2 completes the proof due to Axiom, which we name as when a sequent has the form "$\mathfrak{A} \to \mathfrak{A}$". Then, the proof in Fig. 6 is illustrated in Fig. 7 by using SequentML.

The entire proof is enclosed by <sequentcalculus> tags. Each inference is enclosed by <deduction> tags. The first child of the "deduction" element is the "rule" element, representing an inference rule. The second child is a sequent applied by the inference. If the proof is completed by the inference, the third child is omitted. Else, the "deduction" element would be the third child. This is how a series of sequents in a proof is represented as a nested structure of "deduction" elements.

$$\frac{S2}{S1}$$

Fig. 6 Example of proof

```
<sequentcalculus>
    <deduction>
        <rule>definitions of subset</rule>
        <sequent>S1</sequent>
        <deduction>
            <rule>Axiom</rule>
            <sequent>S2</sequent>
        </deduction>
    </deduction>
</sequentcalculus>
```

Fig. 7 Description of inference using SequentML

In case where a series of sequents has a branch by the inference (Fig. 8), one only has to describe the "deduction" element of S2 (and S3) as the third (and fourth) child of the "deduction" element.

$$\frac{S2 \qquad S3}{S1}$$

Fig. 8 Example of branched sequent series

4 Visualization of Proofs

In this section, we elucidate the visualization of mathematical proofs. When a proof is presented using sequent calculus, it usually comprises a series of sequents. A series of sequents is placed vertically, as shown above. The bottom sequent is the proposition to prove. The repetition of inferences for the relatively simple sequents leads to the completion of the proof if turned sufficiently simple, because of a series of transformations. This vertical placement of sequents is regarded as a type of visual representation. This representation, however, is not appropriate for learners who seek to take advantage of the visualized materials, because the description created using sequent calculus is highly formalized and representation is restrictive regarding the relation between sequents (or inference rules).

In this section, we present a visualization method for overcoming the abovementioned limitations. As an example, we prove De Morgan's laws pertaining to set theory.

4.1 Diagram Structure

The structure of diagrams in a visualized proof is introduced. In sequent calculus, a series of sequents is placed vertically with bars as separators. This makes it difficult to distinguish between the horizontally placed sequents. Further, there is a lack of considerable space for mentioning the relation between the vertically placed sequents. Considering these factors, the authors have devised a new diagram to enclose each sequent within rectangles and to connect the vertically placed sequents using lines. Furthermore, the vertical order of sequents is reversed. This is because in many proofs, steps are taken to gradually simplify the propositions to prove. An example of the visualization of an inference is shown in Fig. 9.

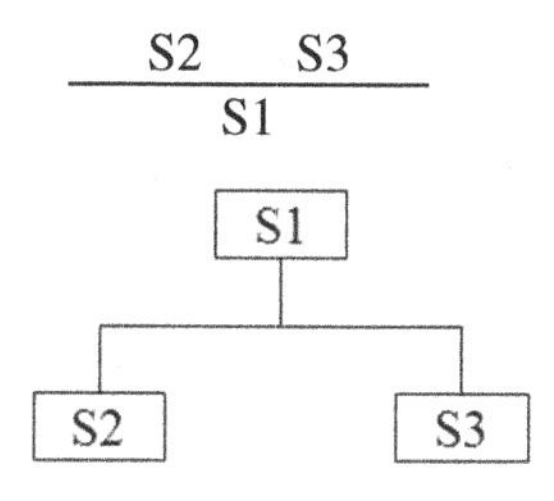

Fig. 9 Example of inference visualization

4.2 Sequents Replaced by Natural Language

As one may tell, most of the general learners of math are probably not familiar with sequent calculus and its style. On the other hand, an aforementioned sequent can be interpreted as a proposition, regarding commas in the antecedent as the logical product, commas in the succedent as the logical disjunction, and an arrow symbol as the logical implication. In the visualization, sequents are interpreted as propositions. However, sequents, regarded as propositions and represented with logical operators, might not be comprehensible since they are considerably formal for learners who have not acquired the basic knowledge of formal logic. Hence, a set of simple rules is devised to replace the logical symbols with the expressions of a natural language. For the corresponding table, see Table 4.

Table 4 Correspondence between logical operatorsand natural language

Logical Operator	Natural Language
$\wedge$	and
$\vee$	or
$\neg$	not
$\Rightarrow$	if… then …
$\Leftrightarrow$	if and only if
$\forall$	for all
$\exists$	there exists

4.3 Representation of Inference Rule

Here, the representation of the relation between sequents, or inference rules, is discussed. A label is attached to the line connecting two sequents, and the content of the “rule” element is directly copied in the label. Fig. 10 shows an example of visualizing an inference connecting S1 and S2 with an inference rule “definition of subset.”

After inferences continue until no further simplification is necessary, a line is drawn below the sequent, implying the end of the inference process. In this occurrence, the reason for why no simplification is required is also provided. The content of the “rule” element of the “deduction” element having no third child is also given. Fig. 11 shows an example of a sufficiently simplified sequent.

Fig. 10 Example of representation of inference rule

Fig. 11 Example of representation of end of inference

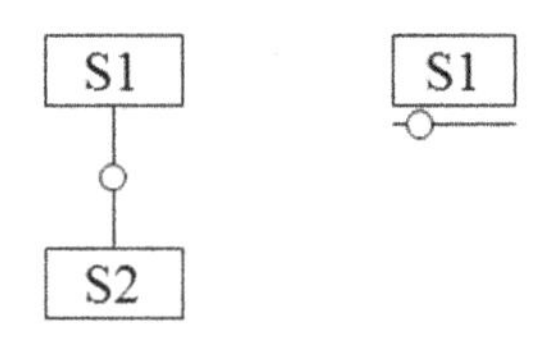

Fig. 12 Abbreviated representation of inference rule

A display of all the inference rules causes the deterioration of the visibility and readability of the logical structure of proofs. In order to avoid this, inference rules are displayed only when a button placed between the sequents is clicked. Fig. 12 shows a status where no inference rules are displayed.

A single inference rule is sometimes used multiple times in a proof. In order to make good use of this rule, a function that allows the referencing of inference rules by using the attributes of the <rule> tag is provided. There are two types of tags to help this function: one is the “id” attribute, and the other is the “href” attribute. The “id” attribute is added to the <rule> tag that describes an inference rule that is used more than once.

```
<rule id="def-subset">definition of subset</rule>
```

When the same inference rule is used, referencing is enabled by specifying the value of the “id” attribute of the <rule> tag with “#” as its prefix, as the value of the “href” attribute.

```
<rule href="#def-subset"/>
```

More generally, one may specify the URI as the “href” attribute. This suggests that the inference rules that are not found in the same file can also be referred. A similar notation is used in HTML + CSS and MathML.

4.4 Case Study

Fig. 15 illustrates an example of a visualized proof. For the sake of comparison, a conventional proof by sequential texts and a diagram for the same proof created by using sequent calculus in which sequents are translated as logical expressions are shown in Fig. 13 and Fig. 14, respectively. The proposition to prove in this case study is the De Morgan's laws belonging to the set theory. One may notice that the proposition to prove is placed at the top, and the propositions are simplified as they get closer to the bottom. The ground for simplification is filled in the label. The sequent at the bottom is sufficiently simplified, and the label for reasoning is attached. It is observed that the visibility and readability are improved by hiding labels. Each sequent is expressed by a natural language on the basis of the aforementioned replacement rule.

Showing $x \in (\overline{A \cup B}) \Leftrightarrow x \in \bar{A} \cap \bar{B}$ is equivalent to showing both $x \in (\overline{A \cup B}) \Rightarrow x \in \bar{A} \cap \bar{B}$ and $x \in \bar{A} \cap \bar{B} \Rightarrow x \in (\overline{A \cup B})$. First, let $x \in (\overline{A \cup B}) \Rightarrow x \in \bar{A} \cap \bar{B}$ be shown. Since $x \in A \cup B$ does not hold, neither $x \in A$ nor $x \in B$ should hold. Hence, both $x \in \bar{A}$ and $x \in \bar{B}$ hold. Therefore, $x \in \bar{A} \cap \bar{B}$. Conversely, the proof for $x \in \bar{A} \cap \bar{B} \Rightarrow x \in (\overline{A \cup B})$ follows. Since neither $x \in A$ nor $x \in B$ holds, $x \in A \cup B$ does not hold. Therefore, $x \in (\overline{A \cup B})$.

Fig. 13 Proof of De Morgan's laws by sequential texts (conventional)

$$\dfrac{\dfrac{\dfrac{\neg(x \in A \vee x \in B) \Rightarrow \neg x \in A \qquad \neg(x \in A \vee x \in B) \Rightarrow \neg x \in B}{\neg(x \in A \ \vee \ x \in B) \Rightarrow \neg x \in A \ \wedge \ \neg x \in B}}{x \in (\overline{A \cup B}) \Rightarrow x \in \bar{A} \cap \bar{B}} \qquad \dfrac{(\neg x \in A \ \wedge \ \neg x \in B) \Rightarrow \neg(x \in A \cup B)}{x \in \bar{A} \cap \bar{B} \Rightarrow x \in (\overline{A \cup B})}}{x \in (\overline{A \cup B}) \Leftrightarrow x \in \bar{A} \cap \bar{B}}$$

Fig. 14 Proof of De Morgan's laws by sequent calculus and its representation

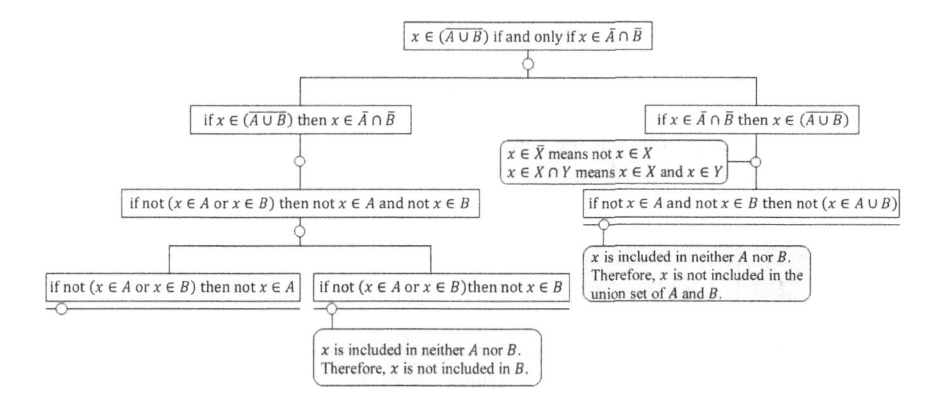

Fig. 15 Visualization of the proof of De Morgan's laws

5 Discussion

This section focuses on the advanced functions of the proposed method for future implementation. The one is, utilizing visualized proofs, a function allowing learners to input proofs by themselves. The other is concerning the similarities of proofs with their structures into considered.

5.1 Describing Proofs by Learners

Learning effectiveness will be enhanced if learners are provided with a tool to input mathematical proofs with this visual representation by themselves. One simple manner in which this can be achieved is to let the inference rules and a set of sequents be null first, and then to let the learners fill them in by using a natural language and have teachers grade them (a fill-in type of problem). On the other hand, if a framework is prepared to let users input formal descriptions in some way, the automatic grading of mathematical proofs will be enabled, taking advantage of the rigid structure by SequentML and MathML. Such an example of automatic verification software for mathematical proofs is MIZAR [7]. A related work is [8], which intended to formalize the nucleus of a MIZAR system using XML. An ideal setting is the automatic grading of natural language-based proofs. [9] presents a research on the analysis of documents which consist of math expressions and a natural language.

5.2 Presenting Similarly Structured Proofs

Another discussion is to consider the similarities between structures of proofs. By defining some index regarding the structure of each proof, it is expected to provide learners with proofs analogous to the ones they studied for better understanding. In addition, it will be made feasible to present the proof of which the learner is more familiar with the structure, in case there are multiple ways to prove the same proposition. In essence, SequentML has a mechanism to describe structures of proofs, sequents, and inference rules. Also, MathML is capable of storing semantic information of math expressions including logical expressions. Therefore, the proofs described by our method should have enough information for their characterization. On the other hand, it is not an easy task to define such an index for judging similarities of proofs, and thorough examination will be indispensable.

6 Concluding Remarks and Future Work

In this paper, the authors developed a framework to create a visual representation of mathematical proofs for math learners. Sequent calculus was adopted as a theoretical platform to describe such proofs. MathML was also used for expressing the math expressions. For the comprehensive data description of sequent calculus, SequentML was designed as an XML vocabulary. Lastly but most importantly, a method was proposed to visualize SequentML-based data from the viewpoint of showing their logical structure clearly.

Our plan is to primarily expand the data stipulated by SequentML and develop a tool to help create such data. In mathematical proofs, identical or similar inference rules are introduced. In this paper, we referred to the transformation rules of sequents as the inference rules, which included the application of theorems or lemmas. Then, the usability of SequentML is expected to be enhanced when the amount of SequentML-based data is increased.

The secondary plan is to add an interactive function for manipulating the visualized diagrams that represent mathematical proofs. In visualized proofs, it is highly expected that the granuarity of the proof is controlled or adjusted depending on the level of the learners. For instance, when some theorem's name is specified as an inference rule, presenting the details of the theorem is desirable if the user is not familiar with the theorem. As its reference, [10] introduced a function to refer to the definitions of mathematical notions or terms when they are clicked. In our method, sequents are displayed by using a natural language at present, while some learners might understand them more easily simply through a description made of logical symbols. A new mechanism will be demanded to vary the diagrams flexibly to the need of each learner. If this system may collaborate with software such as ActiveMath [11], which has learner management functions, it may be feasible to automatically figure out what should be displayed to each learner.

Lastly, the shortcomings of our system have to be revealed. For example, the proof of De Morgan's laws is shown in Fig. 15 in such a way that "if $x \in \overline{(A \cup B)}$ then $x \in \bar{A} \cap \bar{B}$" and "if $x \in \bar{A} \cap \bar{B}$ then $x \in \overline{(A \cup B)}$" are both proved. Another way is to lead to $x \in \bar{A} \cap \bar{B}$ by transforming and replacing the proposition with mathematically equivalent propositions, for a given $x \in \overline{(A \cup B)}$. The latter way, however, cannot be represented by the visualization method presented in this study at present. Then, the next step to take, of course, is to pursue the alternative way to visualize such cases as well.

The implementation will be made shortly based on the presented framework, to observe the effects of the method used in on-site classes.

Acknowledgments. The authors wish to thank the referees for their helpful comments on an earlier version of this paper for deepening the discussions on advanced functions of our method.

References

[1] Hanna, G.: Some pedagogical aspects of proof. Interchange 21(1), 6–13 (1990)
[2] Mariotti, M.A.: Proof and proving in mathematics education. Handbook of Research on the Psychology of Mathematics Education: Past, Present, and Future, pp. 173–204 (2006)

[3] Aczel, J., Fung, P., Bornat, R., Oliver, M., O'Shea, T., Sufrin, B.: Software that assists learning within a complex abstract domain: the use of constraint and consequentiality as learning mechanisms. British Journal of Educational Technology 34(5), 625–638 (2003)

[4] Smullyan, R.M.: First-order logic. Dover Publications, New York (1995)

[5] W3C Math Home, http://www.w3.org/Math/

[6] Gentzen, G.: Untersuchungen über das logische Schlie*β*en. I. Mathematische Zeitschrift 39(1), 176–210 (1935)

[7] Rudnicki, P.: An overview of the Mizar project. In: Proceedings of the 1992 Workshop on Types for Proofs and Programs, pp. 311–330 (1992)

[8] Urban, J.: XML-izing Mizar: making semantic processing and presentation of MML easy. Mathematical Knowledge Management, 346–360 (2006)

[9] Wolska, M., Kruijff-Korbayová, I.: Analysis of mixed natural and symbolic language input in mathematical dialogs. In: Proceedings of the 42nd Annual Meeting on Association for Computational Linguistics, pp. 25–32 (2004)

[10] David, C., Kohlhase, M., Lange, C., Rabe, F., Zhiltsov, N., Zholudev, V.: Publishing math lecture notes as linked data. The Semantic Web: Research and Applications, 370–375 (2010)

[11] Melis, E., Andres, E., Budenbender, J., Frischauf, A., Goduadze, G., Libbrecht, P., et al.: Active-Math: A generic and adaptive web-based learning environment. International Journal of Artificial Intelligence in Education 12, 385–407 (2001)

Incremental Update of Fuzzy Rule-Based Classifiers for Dynamic Problems

Tomoharu Nakashima, Takeshi Sumitani, and Andrzej Bargiela

Abstract. Incremental construction of fuzzy rule-based classifiers is studied in this paper. It is assumed that not all training patterns are given a priori for training classifiers, but are gradually made available over time. It is also assumed the previously available training patterns can not be used in the following time steps. Thus fuzzy rule-based classifiers should be constructed by updating already constructed classifiers using the available training patterns at each time step. Incremental methods are proposed for this type of pattern classification problems. A series of computational experiments are conducted in order to examine the performance of the proposed incremental construction methods of fuzzy rule-based classifiers using a simple artificial pattern classification problem.

1 Introduction

Fuzzy systems based on fuzzy if-then rules have been researched in various fields such as control [1], classification and modeling [2]. A fuzzy rule-based classifier is composed of a set of fuzzy if-then rules. Fuzzy if-then rules are generated from a set of given training patterns. Advantages of fuzzy classifiers are mainly two-folds: First, the classification behavior can be easily understood by human users. This can be done by carefully checking the fuzzy if-then rules in the fuzzy classifier because fuzzy if-then rules are inherently expressed in linguistic forms. Another advantage is nonlinearity in classification. It is well known that non-fuzzy rule-based classifiers

---

Tomoharu Nakashima · Takeshi Sumitani
Graduate School of Engineering, Osaka Prefecture University, Gakuen-cho 1-1, Naka-ku, Sakai, Osaka 599-8531, Japan
e-mail: nakashi@cs.osakafu-u.ac.jp
takeshi.sumitani@ci.cs.osakafu-u.ac.jp

Andrzej Bargiela
University of Nottingham, Jubilee Campus, Wollaton Road, Nottingham NG8 1BB, U.K.
e-mail: andrzej.bargiela@nottingham.ac.uk

R. Lee (Ed.): Computer and Information Science 2012, SCI 429, pp. 209–219.
springerlink.com

are difficult to perform non-linear classification because classification boundaries are always parallel to attribute axes in most cases. The nonlinearity of fuzzy classification leads to high generalization ability of fuzzy rule-based classifiers while its classification behavior is linguistically understood.

A fuzzy rule-based classifier in this paper consists of a set of fuzzy if-then rules. The number of fuzzy if-then rules is determined by the dimensionality of the classification problem and the number of fuzzy partitions used for each attribute. A fuzzy if-then rules is generated by calculating the compatibility of training patterns with its antecedent part for each class. The calculated compatibilities are summed up to finally determine the consequent class of the corresponding fuzzy if-then rule. An unseen pattern is classified by the fuzzy rule-based classifier (i.e. a set of generated fuzzy if-then rules) using a fuzzy inference process.

In general, as the amount of information keeps growing due to the development of high-performance computers and high-capacity memories, it is difficult for any information systems to efficiently and effectively process a huge amount of data at a time. This is because it takes intractably long time to retrieve whole data and it is not possible to handle the intractably huge amount of data by just one information system. Also, it is possible that training patterns are generated over time and the designers of information systems have to handle the dynamically available patterns in a manner of streaming process. This paper focuses on the latter case in the construction process of fuzzy rule-based classifiers. In order to tackle with such streaming data, fuzzy rule-based classifiers need to adapt themselves to newly available training patterns. In this paper, incrementally constructing methods are proposed for fuzzy rule-based classifiers. A series of computational experiments are conducted in order to examine the generalization ability of the constructed fuzzy rule-based classifiers by the proposed methods for a two-dimensional incremental pattern classification problem.

2 Pattern Classification Problems

2.1 Conventional Pattern Classification

The standard type of pattern classification is explained in this subsection. Let us assume that a set of training patterns is given before constructing a classifier. A training pattern consists of a real-valued input vector and its corresponding target class. Consider, for example, an n-dimensional C-class pattern classification problem. It is also assumed that m training patterns $\mathbf{x}_p = (x_{p1}, x_{p2}, \ldots, x_{pn})$, $p = 1, 2, \ldots, m$ are given a priori. The task then is to construct a classifier that correctly classify an unseen pattern using the given set of the training patterns.

2.2 Incremental Pattern Classification

The incremental pattern classification problem in this paper is defined as the classification task that involves an incremental process of obtaining training patterns. That is, the full set of training patterns cannot be obtained beforehand. Instead, a small number of training pattens are gradually made available as the time step proceeds. It is also assumed that classification of new patterns should be made during the course of the incremental process of obtaining training patterns. Thus, a classifier should be constructed using the training patterns that have been made available thus far.

Let us denote the available training patterns at the time step t as $\mathbf{x}_p^t$, $p = 1,2,\ldots,m^t$, where m^t is the number of training patterns that became available at time t. The task at time T is to construct a classifier using $\sum_{t=1}^{T} m^t$ training patterns $\mathbf{x}_p^t$, $p = 1,2,\ldots,m^t, t = 1,2,\ldots,T$.

3 Fuzzy Rule-Based Classifier

In this paper, a fuzzy rule-based classifier proposed in [2] is used. It should be noted that the idea of the classification confidence can be applied to any forms of fuzzy classifiers if they are rule-based systems. An overview of the system in [2] is given below.

3.1 Fuzzy If-Then Rule

In a pattern classification problem with n dimensionality and M classes, we suppose that m labeled patterns, $\mathbf{x}_p = \{x_{p1}, x_{p2}, \cdots, x_{pn}\}$, $p = 1,2,\cdots,m$, are given as training patterns. We also assume that without loss of generality, each attribute of $\mathbf{x}_p$ is normalized to a unit interval $[0,1]$. From the training patterns we generate fuzzy if-then rules of the following type:

$$\begin{aligned} R_q\text{: If } x_1 \text{ is } A_{q1} \text{ and } \cdots \text{ and } x_n \text{ is } A_{qn} \\ \text{then Class } C_q \text{ with } CF_q, \\ q = 1,2,\cdots,N, \end{aligned} \tag{1}$$

where R_q is the label of the q-th fuzzy if-then rule, $\mathbf{A}_q = (A_{q1}, \cdots, A_{qn})$ represents a set of antecedent fuzzy sets, C_q a the consequent class, CF_q is the confidence of the rule R_q, and N is the total number of generated fuzzy if-then rules.

We use triangular membership functions as antecedent fuzzy sets. Figure 1 shows triangular membership functions which divide the attribute axis into five fuzzy sets. Suppose that an attribute axis is divided into L fuzzy sets. The membership function of the k-th fuzzy set is defined as follows:

$$\mu_k(x) = \max\left\{1 - \frac{|x - x_k|}{v}, 0\right\}, k = 1,\cdots,L, \tag{2}$$

where

$$x_k = \frac{k-1}{L-1}, \; k = 1,\cdots,L, \tag{3}$$

and

$$v = \frac{1}{L-1}. \tag{4}$$

Let us denote the compatibility of a training pattern $\mathbf{x}_p$ with a fuzzy if-then rule R_q as $\mu_{\mathbf{A}_q}(\mathbf{x}_p)$. The compatibility $\mu_{\mathbf{A}_q}(\mathbf{x}_p)$ is calculated as follows:

$$\mu_{\mathbf{A}_q}(\mathbf{x}_p) = \prod_{i=1}^{n} \mu_{A_{qi}}(x_{pi}), \; q = 1, 2, \cdots, N, \tag{5}$$

where $\mu_{A_{qi}}(x_{pi})$ is the compatibility of x_{pi} with the fuzzy set A_{qi} and x_{pi} is the i-th attribute value of $\mathbf{x}_p$. Note that $\mu_{A_{qi}}(x_{pi})$ is calculated by (2).

The number of fuzzy rules to be generated is L^n. That is, the number of rules increases exponentially for the division number and the dimensionality.

3.2 Generating Fuzzy If-Then Rules

A fuzzy classification system consists of a set of fuzzy if-then rules. The fuzzy if-then rules are generated from the training patterns $\mathbf{x}_p$, $p = 1, 2, \ldots, m$. The number of generated fuzzy if-then rules is determined by the number of fuzzy partitions for each axis (i.e., L in (2) $\sim$ (4)). That is, the number of generated fuzzy if-then rules is the number of combinations of fuzzy sets that are used for attribute axes. Although different numbers of fuzzy partitions can be used for different axes, in this paper we assume that it is the same for all axes. In this case, the number of fuzzy if-then rules is calculated as $N = L^n$ where n is the dimensionality of the pattern classification problem at hand. In this paper, it is supposed that all attributes are divided in the same way (i.e., the same fuzzy partition).

The consequent part of fuzzy if-then rules (i.e., C_q and CF_q in (1)) is determined from the given training patterns once the antecedent part is specified. The consequent class C_q of the fuzzy if-then rule R_q is determined as follows:

$$C_q = \arg \max_{h=1,\ldots,M} \beta_h^q, \tag{6}$$

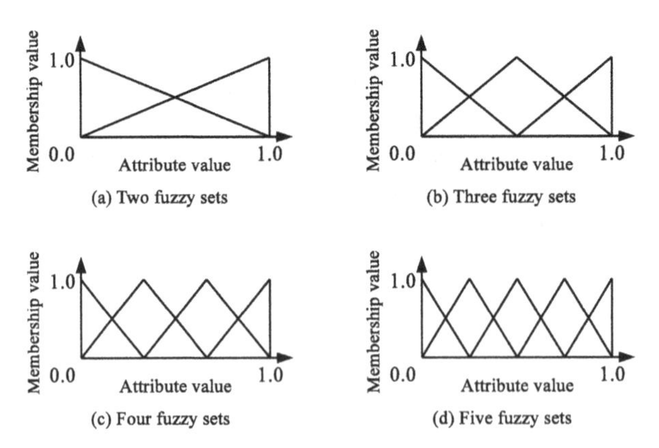

Fig. 1 Triangular fuzzy sets.

where

$$\beta_h^q = \sum_{\mathbf{x}_p \in \text{Class } h} \mu_{\mathbf{A}_q}(\mathbf{x}_p). \tag{7}$$

That is, the most matching class with the fuzzy if-then rule is selected considering the given training patterns. If there is not any training pattern that is covered by the fuzzy if-then rules, the consequent class is set as empty. Also, in the case where multiple classes have the maximum value in (6), the consequent class is set as empty. The confidence CF_q is determined as follows:

$$CF_q = \frac{\beta_{C_q} - \bar{\beta}}{\sum_{h=1}^{m} \beta_h^q}, \tag{8}$$

where

$$\bar{\beta} = \frac{1}{M-1} \sum_{h \neq C_q} \beta_h^q. \tag{9}$$

There are other formulations for determining the confidence. Interested readers are referred to [2] for the discussion on the confidence calculation and the performance evaluation.

3.3 *Classification of Unseen Patterns*

Generated fuzzy if-then rules in the previous subsections are used to assign a class label to an unseen pattern which is not included in the set of training patterns. Let us denote an n-dimensional unseen pattern as $\mathbf{x} = (x_1, x_2, \ldots, x_n)$. The fuzzy inference is employed to classify unseen patterns in the fuzzy classification system in this paper. The class of an unseen pattern $\mathbf{x}$ is classified as Class C that is determined by the following equation:

$$C = \arg \max_{h=1,\ldots,M} \{\alpha_h\}, \tag{10}$$

where

$$\alpha_h = \max_{\substack{q=1,\ldots,N \\ C_q=h}} \{\mu_{\mathbf{A}_q(\mathbf{x})} \cdot CF_q\}. \tag{11}$$

In the above equations, M is the number of classes and N is the number of fuzzy if-then rules. In (10), if there are multiple classes that have the same maximum value of α_h, the classification of the unseen pattern is rejected.

4 Incremental Construction of Fuzzy Rule-Based Classifiers

Since not all training patterns are available at a time but are available over time, it is necessary for already generated fuzzy if-then rules to adapt themselves to the training patterns that are newly made available. In this paper three methods for incrementally constructing fuzzy rule-based classifiers are proposed. The three methods update the summed compatibilities that are calculated in (7), but in different manners.

Incremental method A: The summed compatibilities are updated so that the new training patterns are considered equally as the previously available training patterns.
Incremental method B: The summed compatibilities are updated so that the higher weights are put for the new training patterns than the previously available training patterns.
Incremental method C: The summed compatibilities are updated so that

The following subsections explain the above incremental construction methods of fuzzy rule-based classifiers.

4.1 Incremental Method A

Let us assume that at time T a fuzzy classifier has been already constructed. It is also assumed that at time $T+1$ new training patterns $\mathbf{x}_p^{T+1}$, $p = 1,2,\ldots,m^{T+1}$ are made available. As in Subsection 3.2, each fuzzy if-then rule has a summed compatibility β for each class.

The update procedure of the summed compatibilities is written as follows:

For each of new training patterns $\mathbf{x}_p^{T+1}$, $p = 1,2,\ldots,m^{T+1}$, do the following steps:

Step 1: Calculate the compatibility $\mu_q(\mathbf{x}_p^{T+1})$ of $\mathbf{x}_p^{T+1}$ with the j-th fuzzy if-then rule R_q, $q = 1,2,\ldots,N$, where N is the total number of generated fuzzy if-then rules.
Step 2: Update the summed compatibilities β_h^q of the fuzzy if-then rule R_q for Class h as follows:

$$\beta_h^q \leftarrow \begin{cases} \dfrac{n_h^q \cdot \beta_h^q + \mu_q(\mathbf{x}_p^{T+1})}{n_h^q + 1} & \text{, if } \mathbf{x}_p^{T+1} \in \text{Class } h \\ & \text{and } \mu_q(\mathbf{x}_p^{T+1}) > 0.0, \\ \beta_h^q & \text{, otherwise.} \end{cases} \quad (12)$$

The above procedure calculates the new summed compatibility as the weighted average of the previous summed compatibility and the compatibility with the new pattern. The weight assigned for the summed compatibility is the number of training patterns that were previously available. Thus the new summed compatibility can be seen as the summed compatibility for $\mathbf{x}_p$, $p = 1,2,\ldots,m^t$, $t = 1,2,\ldots,T+1$ that is calculated by the standard fuzzy rule-generation procedure in Subsection 3.2.

4.2 Incremental Method B

The second method of incrementally constructing fuzzy rule-based classifiers does not take into consideration of the number of training patterns that were used to calculate the summed compatibilities. Instead, they are modified so that they approach the compatibility of the new training patterns. The following steps explain the procedure:

For each of new training patterns $\mathbf{x}_p^{T+1}$, $p = 1, 2, \ldots, m^{T+1}$, do the following steps:

Step 1: Calculate the compatibility $\mu_q(\mathbf{x}_p^{T+1})$ of $\mathbf{x}_p^{T+1}$ with the j-th fuzzy if-then rule R_q, $q = 1, 2, \ldots, N$, where N is the total number of generated fuzzy if-then rules.

Step 2: Update the summed compatibilities β_h^q of the fuzzy if-then rule R_q for Class h as follows:

$$\beta_h^q \leftarrow \begin{cases} \beta_h^q + \gamma \cdot \delta_h^q, & \text{if } \mathbf{x}_p^{T+1} \in \text{Class } h \\ & \text{and } \mu_q(\mathbf{x}_p^{T+1}) > 0.0, \\ \beta_h^q, & \text{otherwise,} \end{cases} \tag{13}$$

where

$$\delta_h^q = \sum_{\mathbf{x}_p \in \text{Class } h} \mu_{\mathbf{A}_q}(\mathbf{x}_p) - \beta_h^q, \tag{14}$$

and γ is a positive constant in the closed interval $[0.0, 1.0]$.

4.3 Incremental Method C

The third incremental method considers the previously available training patterns with decayed weights. In this method, the value β for each class and each fuzzy if-then rule at time step T is calculated instead of (7) as follows:

$$\beta_h^{q,T} = \sum_{t=1}^{T} \sum_{\mathbf{x}_p \in \text{Class } h} \mu_{\mathbf{A}_q}(\mathbf{x}_p) \cdot \gamma^{T-t}, \quad h = 1, 2, \ldots, C, \quad q = 1, 2, \ldots, N, \tag{15}$$

where γ is a positive constant called a decay rate in the range $[0.0, 1.0]$. In this equation, the latest training patterns available at time step T are not influenced by the decay rate (i.e., $\gamma^{T-T} = 1.0$). The older the training patterns made available, the weaker their weight becomes (i.e., γ^{T-1} for the oldest training patterns that are available at time step 1).

When new patterns are available, the value of β is updated using (15). Since the calculation for the beta value in (15) is made for all the training patterns that are made available so far, the computational cost for updating β's is intractably high. Let us consider now the value of β for time step $T + 1$ as follows:

$$\begin{aligned}
\beta_h^{q,T+1} &= \sum_{t=1}^{T+1} \sum_{\mathbf{x}_p \in \text{Class } h} \mu_{\mathbf{A}_q}(\mathbf{x}_p) \cdot \gamma^{T+1-t} \\
&= \sum_{t=1}^{T} \sum_{\mathbf{x}_p \in \text{Class } h} \mu_{\mathbf{A}_q}(\mathbf{x}_p) \cdot \gamma^{T+1-t} \\
&\quad + \sum_{\mathbf{x}_p \in \text{Class } h} \mu_{\mathbf{A}_q}(\mathbf{x}_p) \cdot \gamma^{T+1-(T+1)} \\
&= \gamma \cdot \sum_{t=1}^{T} \sum_{\mathbf{x}_p \in \text{Class } h} \mu_{\mathbf{A}_q}(\mathbf{x}_p) \cdot \gamma^{T-t} \\
&\quad + \sum_{\mathbf{x}_p \in \text{Class } h} \mu_{\mathbf{A}_q}(\mathbf{x}_p) \\
&= \gamma \cdot \beta_h^{q,T} + \sum_{\mathbf{x}_p \in \text{Class } h} \mu_{\mathbf{A}_q}(\mathbf{x}_p). \qquad (16)
\end{aligned}$$

That is, it is not necessary to calculate the value of β from scratch but the previous β can be used as a base for calculating the new vlaue of β. The new value of β is calculated as the weighted sum of the previous β and the compatibility values for new training patterns.

5 Computational Experiments

The performance of the proposed methods were examined for a static incremental pattern classification problem. During the course of the experiments, the classification boundaries remain unchanged. Figure 2 shows the problem that is used in this subsection. A training pattern in this classification problem constitutes of a two-dimensional input vector in the domain space $[0.0, 1.0]^2$ and a target output which is one of four classes. The two diagonal lines in the two-dimensional pattern space are the classification boundaries between the four classes.

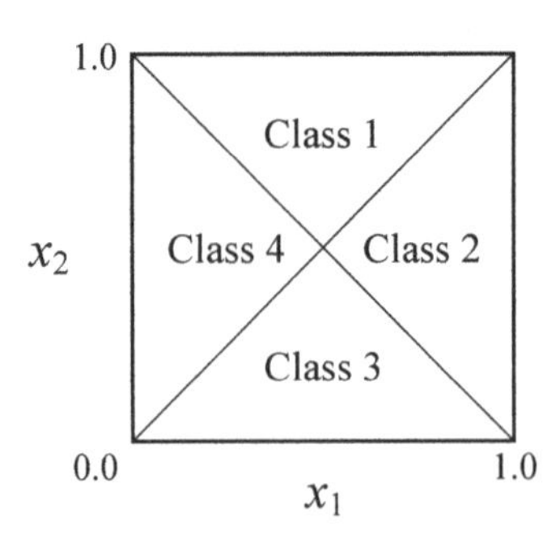

Fig. 2 A static pattern classification problem.

A fuzzy rule-based classifier is constructed from training patterns that are temporarily available. In the computational experiments, the number of temporarily available training patterns at a single time unit is specified as 1, 5, 10, and 100. The new training patterns are generated by first randomly sampling a two-dimensional input vector according to the uniform distribution, then specify the target output as the class in which the sampled vector falls. The procedure of generating new training patterns was iterated for 100 times. Thus the total number of new training patterns in a single run is 100, 500, 1000, and 10000 depending on the number of generated training patterns at a single time step. For the purpose of examining the generalization ability of the constructed fuzzy if-then rules, 10000 input patterns are uniformly generated in the domain space $[0,1]^2$. The target class of a test pattern is determined by the area in which the pattern falls. For Incremental method B, the value of the positive constant γ in (13) is specified as $0.0, 0.1, \ldots, 0.9$, and 1.0. The number of fuzzy sets for each attribute is specified as two, three, four, and five (see Fig. 1). For each parameter specification, the computational experiments were conducted ten times to obtain the average performance of the proposed methods.

The experimental results by Incremental method A is shown in Table 1. From this table, it is shown that the classification performance becomes better when the number of available training patterns is larger, which is quite natural as the training classifiers is effective with more training patterns. The classification performance is not very good when the number of fuzzy sets is two and four. This is because the current setting of the triangular fuzzy partitioning is not suitable for the diagonal classification boundaries when the number of fuzzy sets is even (not odd). As Incremental method A eventually considers all training patterns equally at the end of the process. This method should be better than Incremental method B which updates fuzzy if-then rules in a smoothing manner with a positive constant γ.

Table 1 Experimental results by Incremental method A for the static problem. The number in each entry shows the classification performance (%).

	# of patterns				
# of fuzzy sets	1	2	5	10	100
2	48.0	49.7	49.5	50.3	49.5
3	91.8	92.7	94.4	94.9	95.6
4	82.3	82.9	84.3	84.7	95.4
5	90.9	92.2	94.1	94.3	95.3

The results of the experiments by Incremental method B are shown in Fig. 3. Figure 3 shows the classification performance of the constructed fuzzy rule-based classifiers over time with the specified number of available training patterns and a positive constant (i.e., γ). Obviously the classification performance by Incremental method A is better than Incremental method B as the classification problem is static, that is, they remain unchanged during the course of the experiments. As for the specification of the positive constant, the classification performance is the best when $\gamma = 0.1$ among the investigated values for any number of fuzzy partitions except two.

As shown in the experiments for Incremental method A, it is difficult to achieve the diagonal classification boundaries by two fuzzy sets for each axis. Thus the classification performance in this parameter specification is rather unstable in the low classification rates.

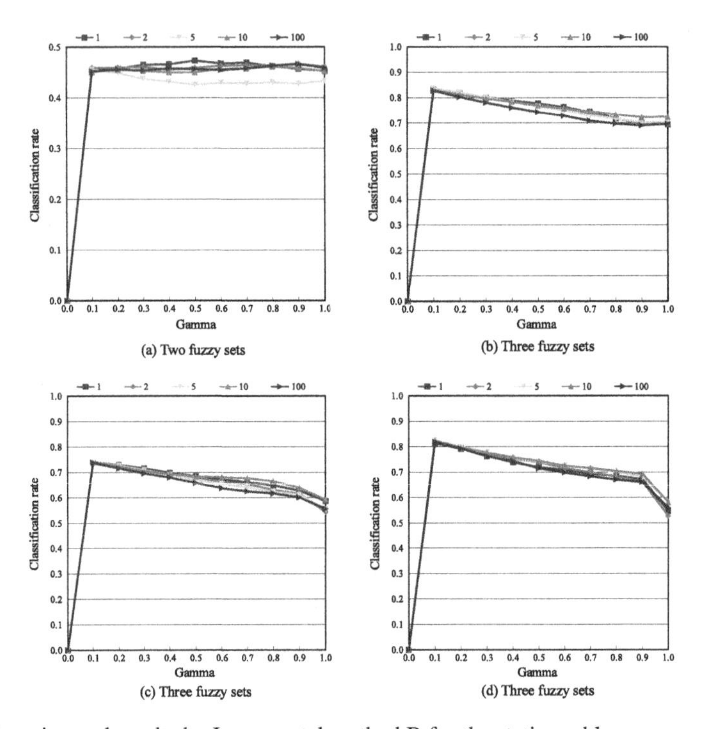

Fig. 3 Experimental results by Incremental method B for the static problem.

Figure 4 shows the results by Incremental method C where training patterns are weighted so that old training patterns do not have much influence on the determination of the consequent class of fuzzy if-then rules. The similar discussion can be made as to the case in Incremental method B. That is, when the number of fuzzy partitions is two for each axis, the performance of the fuzzy rule-based systems is not high dut to the limitation of classification ability with the coarse fuzzy partitions for pattern classification with diagonal boundaries. For Incremental method C, the classification performance increases as the value of γ increases. This is because the classification problem is not dynamic but static. In this classification problem, the information on all available training patterns should be preserved as they remain true at any time steps. Once the classification problem becomes dynamic (i.e., the classification boundaries changes over time), the large value of γ would not be useful. It would be necessary to find the optimal value for dynamic problems.

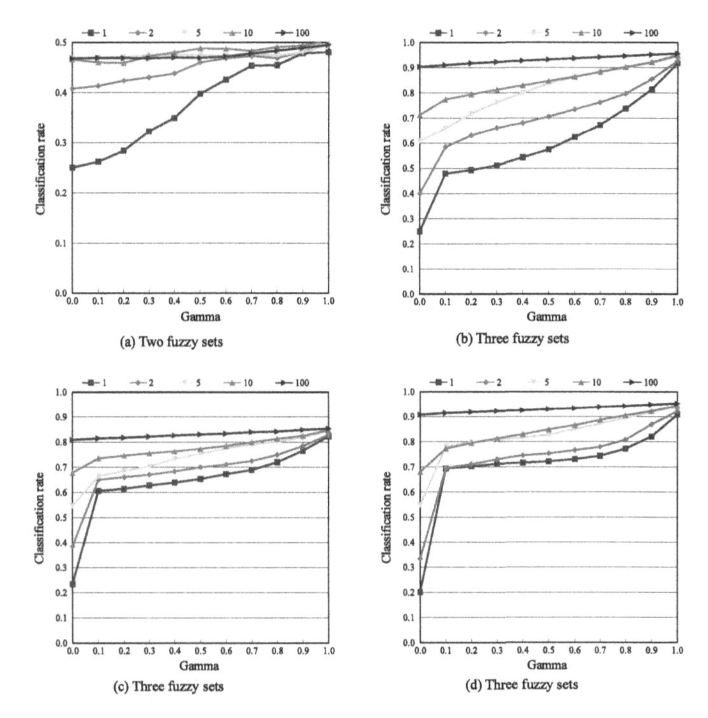

Fig. 4 Experimental results by Incremental method B for the static problem.

6 Conclusions

Two incremental versions of fuzzy rule-based classifiers was proposed in this paper: Incremental method A and B. The target classification problems are supposed to give streaming training patterns that are made available over time. Fuzzy if-then rules have to be updated according to the new training patterns. Through a series of computational experiments, the performance of the proposed fuzzy rule-based classifiers was examined by two types of classification problems: static and dynamic. The experimental results showed that Incremental method A performs better than Incremental method B for the static problem. The analysis of computational cost for the incremental learning should be also considered for the implementation of the methods in real-world applications. The application of the proposed method to real-world problems are also considered in our future works.

References

1. Lee, C.C.: Fuzzy logic in control systems: Fuzzy logic controller – part I and II. IEEE Transactions on Systems Man and Cybernetics 20(2), 404–435 (1990)
2. Ishibuchi, H., Nakashima, T., Nii, M.: Classification and Modeling with Linguistic Information Granules: Advanced Approaches to Linguistic Data Mining, 1st edn. Springer New York, Inc., Secaucus (2004)

Author Index